Texts in Philosophy
Volume 6

A Realist Philosophy of Mathematics

Volume 1
Knowledge and Belief
Jaakko Hintikka

Volume 2
Probability and Inference: Essays in Honour of Henry E. Kyburg
Bill Harper and Greg Wheeler, eds

Volume 3
Monsters and Philosophy
Charles T. Wolfe, ed.

Volume 4
Computing, Philosophy and Cognition
Lorenzo Magnani and Riccardo Dossena, eds

Volume 5
Causality and Probability in the Sciences
Federica Russo and Jon Williamson, eds

Volume 6
A Realist Philosophy of Mathematics
Gianluigi Oliveri

Texts in Philosophy Series Editors
Vincent F. Hendricks   vincent@ruc.dk
John Symons   jsymons@utep.edu

A Realist Philosophy
of Mathematics

by

Gianluigi Oliveri

© Individual author and College Publications 2007 All rights reserved.

ISBN 978-1-904987-40-6
Published by College Publications
Scientific Directors: Dov Gabbay, Vincent F. Hendricks and John Symons
Managing Director: Jane Spurr
Department of Computer Science
King's College London
Strand, London WC2R 2LS, UK

http://www.collegepublications.co.uk

Original cover design by Richard Fraser,
Adapted and prepared by Orchid Creative, www.orchidcreative.co.uk
Printed by Lightning Source, Milton Keynes, UK

All rights reserved. No part of this publication may be reproduced, stored in a retrieval system or transmitted, in any form, or by any means, electronic, mechanical, photocopying, recording or otherwise, without prior permission, in writing, from the publisher.

CONTENTS

PREFACE ix

INTRODUCTION 1

CHAPTER 1 THE REALISM/ANTI-REALISM DEBATE IN THE PHILOSOPHY OF MATHEMATICS: SOME QUESTIONS ABOUT METHODOLOGY 3
1 Introduction 3
2 A main task of the philosophy of mathematics 3
3 The centrality of the question about realism 5
4 The problem of the framework 7
5 The myth of the framework 9
6 A case-history: the infinitesimals 11
7 From logic to metaphysics? 17
8 Assessing some Dummettian claims 20
9 A case in favour of independence 26
10 A new form of debate? 29
11 An appeal to history 31
12 Naturalism about mathematics? 36

CHAPTER 2 ARGUMENTS FOR REALISM IN MATHEMATICS 39
1 Introduction 39
2 From realism in mathematics to set-theoretical realism 40
3 The objectivity argument 41
4 Two case-histories 43
5 The identity and inexhaustibility arguments 49
6 The theory acceptability argument 52
7 Some case-histories 53
8 The faithful representations argument 59
9 The indispensability thesis 61
10 Field and the Indispensability Thesis 63
11 Critical remarks 65
12 More against the Indispensability Thesis 73

CHAPTER 3   TRADITIONAL REALIST POSITIONS   79
1 Introduction 79
2 Pythagorean realism 80
3 Aristotelian realism 82
4 Kantian realism 86
5 Platonism 97
6 Platonism about objects 98
7 Fregean platonism 100
8 Mathematics as an eidetic science 105
9 Mathematics as a science of concepts 111
10 Platonism about structures 117
11 Platonism about structures, its importance 119
12 Platonism about structures: some objections 121

CHAPTER 4   ANTI-REALISM IN MATHEMATICS   127
1 Introduction 127
2 Formalism and Hilbert's programme 127
3 Formalism: a critical assessment 130
4 Nominalism in mathematics 134
5 Fictionalism in mathematics 135
6 Field's anti-objectivism 137
7 Constructivism 139
8 Brouwer's Intuitionism 142
9 Dummett's Intuitionism 148
10 Markov's constructivism 152
11 Bishop's philosophy of mathematics 155
12 Strict finitism 158

CHAPTER 5   THE THIRD WAY: A REALISM WITH THE HUMAN FACE   163
1 Introduction 163
2 Patterns 163
3 Objects and properties of objects 165
4 External and internal relations 166
5 Neither *in re* nor *ante rem* structures 171
6 Which realism? 173
7 Kant on concepts and perception 176
8 Seeing or interpreting? 179
9 Perceptions and expectations 182
10 Psychologism? 184
11 Two objections 186

12	Sets, patterns, and infinity	190

CHAPTER 6 MATHEMATICS: A QUASI-EMPIRICAL SCIENCE? 195

1	Introduction	195
2	A basic distinction	196
3	Anti-empiricism in mathematics	196
4	Is mathematics an empirical science?	197
5	The empiricist's dilemma	200
6	Quasi-empiricism	202
7	Is mathematics quasi-empirical?	204
8	Euclidean geometry: a case-history (I)	206
9	Euclidean geometry: a case-history (II)	210
10	Scientific Research Programmes	214
11	Mathematical Research Programmes	218
12	Quasi-empiricism, foundations, foundationalism	220

CHAPTER 7 A RATIONAL RECONSTRUCTION OF CANTOR-ZERMELO SET THEORY 223

1	Introduction	223
2	The pre-history: Cauchy analysis	225
3	The pre-history: the theory of functions	226
4	The hard core of a new MRP (1)	227
5	The hard core of a new MRP (2)	230
6	The hard core of a new MRP (3)	233
7	The Cantor-Zermelo MRP: the protective belt	237
8	Zermelo's system Z: the hard core	240
9	The protective belt of Z	242
10	Problems with Z, and the system ZFC	246
11	Anti-realist rivals of the Cantor-Zermelo MRP	250
12	A new argument for realism	253

APPENDIX I THE BHK-INTERPRETATION 257

BIBLIOGRAPHY 258

INDEX 267

PREFACE

The Philosophy of Mathematics is at the present moment in an intriguing but confused state. An interesting account of the situation is to be found in [Zheng, 1997]. Zheng begins with the foundationalist schools in the philosophy of mathematics which flourished in the period 1890 to 1940, that is to say with logicism, intuitionism and formalism. Of course these schools continue to some extent to this day, since there are neo-logicists, neo-intuitionists (or neo-constructivists), and neo-formalists. Still, these approaches no longer represent the bulk of the research in the subject, and I think that Zheng is right to say (p. 158) that 'the foundationalist epoch of the philosophy of mathematics has now passed'.

But what has replaced the foundationalist epoch? Zheng discerns two main approaches to the subject which are, however, so different as to be virtually non-overlapping. The first of these could be called the *historical* approach. It goes back to [Lakatos, 1963-4]. Its main characteristic is to relate philosophy of mathematics to examples from the history of mathematics in the way that is done in the history and philosophy of science. [Tymoczko, 1986], [Aspray & Kitcher, 1988], [Gillies, 1995], and [Grosholz & Breger, 2000] are examples of edited collections of papers which adopt this approach. Interestingly, philosophers of mathematics of this school are not exclusively taken up with past history but often try to relate their philosophy to contemporary mathematical practice. A recent example of this is [Corfield 2003,].

Let us now turn to the other approach to the subject. This characteristically ignores both the history of mathematics and the contemporary mathematical scene in favour of traditional philosophical problems concerned with the ontology and epistemology of mathematics. The central question for philosophers who adopt this approach is whether numbers exist, and, if so, in what sense. Zheng refers to this approach (p. 165) as 'philosophy of mathematics for philosophers'. I will call it the *analytic* approach to the philosophy of mathematics since it is adopted mainly by contemporary analytic philosophers.

At present these two approaches hardly seem to interact as Zheng shows in the following striking passage (p. 165):

> The difference between the two different paradigms or different directions in the recent development of the philosophy of mathematics can be seen very clearly by comparing some particularly authorized books in this field. For example, it is quite interesting to compare the contents of the following two books: (1) *New Directions in the Philosophy of Mathematics*, edited by T. Tymoczko [1986], which by its title, can be recognized immediately as belonging to the revolutionary side; and (2) *The Philosophy of Mathematics*, edited by W. Hart [1998], which I think should be classified as belonging to the other side. As a matter of fact, although both the editors claimed that their book included the most interesting and important work from the recent years in the philosophy of mathematics, or at least 'get (got) together some of the most exciting essays published recently in this field', there is not even one common paper in these two books. It proves that they are really two different paradigms, or we might say, these two editors did seem to live in two different worlds.

Given then that there are two different and non-interacting approaches to the philosophy of mathematics, one does not need to be a Hegelian to think that progress might be achieved by forming a synthesis between the two. The interest and originality of Oliveri's new book lies precisely here, because his book is one of the first to present a coherent synthesis of the two approaches.

Oliveri is well-qualified to attempt this task. He wrote his D.Phil dissertation at Oxford on *The principles of analytical philosophy* under the supervision of Professor Sir Michael Dummett. This is certainly a strong background for the analytic approach to the philosophy of mathematics. However, Oliveri is also a follower of Imre Lakatos, the founding father of the historical approach to the philosophy of mathematics. In fact it is noticeable that his synthesis is strongly influenced by this bringing together of Dummett and Lakatos.

The first four chapters of Oliveri's book very much belong to the analytic approach. They deal with the realism/anti-realism debate in the philosophy of mathematics which, as already remarked, is perhaps the central problem of the analytic approach. Here the reader will find a very good critical account of the various positions from the Pythagoreans to Dummett and Field. There is, however, one historical note which would not normally be found in an analytic treatment. This is in chapter 1, section 6 which gives a case-history — the infinitesimals. The problem with the question as to whether mathematical entities exist is that it might be dismissed by the working mathematician as a purely philosophical question which has little bearing on mathematical practice. However, Oliveri's historical example of the infinitesimals shows that this need not be the case. On the contrary the question of whether infinitesimals exist had considerable implications for mathematical practice and led to the development of many important mathematical theories. The case of infinitesimals also shows, according to Oliveri, that questions of existence can be (p. 15): 'dealt with by the mathematical community through the use of rational criteria'.

In chapter 5, Oliveri presents his own version of realism according to which (p. 163) 'mathematics is a science of patterns, where patterns are neither objects nor properties of objects, but aspects, or aspects of aspects, etc. of concrete objects'. The use of the term 'aspect' here contains an implicit reference to Part II of Wittgenstein's *Philosophical Investigations*. It is here that Wittgenstein gives his famous example of the duck-rabbit. Seeing this figure as a rabbit is, according to Wittgenstein, seeing an aspect, and Wittgenstein remarks ([Wittgenstein, 1983], Part II, p. 212^e) that 'what I perceive in the dawning of an aspect is not a property of the object'.

Oliveri's application of the Wittgensteinian concept of aspect to the philosophy of mathematics is a novel development, but it is the last two chapters of the book (6 and 7) which contain the really new twist. It is in these two chapters that Oliveri introduces the ideas of Lakatos and applies them to the realism problem — a problem which Lakatos himself never considered.

Lakatos developed his methodology of scientific research programmes as a way of analysing science, and he was planning to apply this approach to mathematics when his life was cut short at an early age. Oliveri takes up this approach and develops a notion of mathematical research programme (MRP) which he applies to two examples: Euclidean geometry and Cantor-Zermelo set theory. Cantor-Zermelo set theory has, according to Oliveri proved to be a progressive mathematical research programme, whereas constructivism (or intuitionism) has been less successful. Indeed this latter framework theory has not yet reached the stage of being a well-defined mathematical research programme at all. Thus, as things stand at present, it is rational to prefer Cantor-Zermelo set theory to constructivism (or intuitionism). It is at this point that the connection with the realism problem is made, for, according to Oliveri (p. 253): 'accepting an MRP is an act that commits one to accepting the realist or anti-realist metaphysical assumptions contained in its hard core. Therefore, if we, in particular, accept Cantor-Zermelo set theory, we commit ourselves to a realist view of set theory, and of the mathematical theories unified by it'. Oliveri describes this as 'a new argument for realism', and he is right to do so. It is one of the several new and interesting things in this book which mean that it can be strongly recommended to the philosophy of mathematics community.

Donald Gillies
University College London

INTRODUCTION

One of the most important tensions which have traditionally animated the philosophical debate about mathematics is that between the philosophers who believe that mathematics is a science that produces information about some kind of reality, and those who do not.

As is well known, after more than 2500 years of mathematical practice and philosophical reflection on mathematics, we appear to be no nearer to relaxing this tension, in the sense that after all this time, and all the effort that has gone into the debate between realists and anti-realists about mathematics, we are no nearer to reaching an agreement, among the philosophers involved in the controversy, on the nature of this subject, a subject which is so central to our culture and the way we relate to the world.

Another profound and unresolved controversy in the philosophy of mathematics is that on the existence and nature of mathematical knowledge. It is a fact that Logicism, Intuitionism, and Hilbert's programme did not succeed in their intent of providing mathematics with unshakeable foundations; and that what they left in their wake, besides many contributions to mathematics and to the clarification of important philosophical concepts, was a wary scepticism regarding the possibility of reaching any objective and compelling conclusion concerning the existence and nature of mathematical knowledge.

If we, now, consider mathematics in the light of the observations above, we cannot help being struck — like Russell once was — by the scandal that, with regard to such an important and celebrated subject, there is neither agreement concerning what it is about, if anything at all, nor there is any agreement on whether it produces knowledge and, eventually, on which kind of knowledge this is.

The main aim of this book is resolving what I have called the scandal of the philosophy of mathematics. And the way I have chosen to do this consists of two steps.

First, presenting and defending a non-Platonist form of mathematical structural realism which, in the respect of the history of mathematics, harmonizes with a plausible epistemology that naturally arises from it.

Secondly, arguing that, in contrast with the traditional foundationalist approach to these matters proper to Logicism, Intuitionism, and Hilbert's programme, mathematics is a quasi-empirical science, i.e., that, although the process of mathematical justification (proof) has an *a priori* nature, in mathematics, like in the empirical sciences, the concept of fallibility plays a central rôle.

Given the complexity of the problems at hand, and the plurality of the approaches to these problems present in the literature, the book opens — Chapter 1 — with an analysis of the method that should be adopted to deal with the realism/anti-realism debate in the philosophy of mathematics.

An account is, then, given — Chapters 2–4 — of the present state of the art in the debate between realists and anti-realists in the philosophy of mathematics.

The remaining part of the book is dedicated to the defense of the new realist view of mathematics I mentioned above, and of the idea that mathematics is a quasi-empirical science.

Many people have helped me, in one way or another, during the long process of writing this book. And, therefore, before bringing this introduction to a close, I wish to thank them all for all they have done. In particular, I would like to mention here Michael Dummett, Donald Gillies, Enrico Martino, Alessio Plebe, and Gianni Rigamonti with whom I have had the privilege of discussing my work in detail; my colleagues in the Philosophy Department at the University of Palermo — especially Franco Lo Piparo e Marco Carapezza — for having allowed me to have, for two consecutive years, a whole semester free from teaching; the Department of Cognitive Science of the University of Messina, for having awarded me a doctorate in Cognitive Science for a dissertation based on this book; Ms. Hilla Wait, and her staff, at the Philosophy Faculty Library of the University of Oxford, for their helpfulness; and the Fellows of Wolfson College, Oxford, where it all began.

Palermo
May 2006

GO

CHAPTER 1

THE REALISM/ANTI-REALISM DEBATE IN THE PHILOSOPHY OF MATHEMATICS: SOME QUESTIONS ABOUT METHODOLOGY

1 Introduction

The realism/anti-realism debate is one of the traditional central themes in the philosophy of mathematics. The controversies about the existence of the irrational numbers, the complex numbers, the infinitesimals, etc. will be familiar to all who are acquainted with the history of mathematics.

However, in spite of the several attempts to resolve it, attempts made by some of the sharpest minds working in the subject, the realism/anti-realism debate about mathematics has remained very much an open sore to the present day.

This situation calls, therefore, for much caution in venturing any contribution to the discussion. And it is caution what suggests to engage, in this first chapter, in a preliminary survey and critical evaluation of some of the most important, and still influential, methodological suggestions on how to formulate the realism/anti-realism debate in the philosophy of mathematics.

2 A main task of the philosophy of mathematics

From the beginning, one of the main tasks of philosophical investigation has been that of producing overviews of particular areas of human interest. Philosophy has always striven to discover the right way of looking at the world or at particular aspects of it. For instance, given that there are important knowledge-producing activities which we call 'scientific', much of the philosopher's attention directed to them has been aimed at gaining an overview of such activities through the attempt to determine what knowledge is.

It is important to realise how different this type of research is from that in which are engaged those who are involved in developing the sciences. In fact, whereas physicists, biologists, etc. describe and explain natural phenomena; what is ultimately produced by an investigation about knowledge is, instead, understanding of what a scientific theory is; and such an understanding is the consequence of a conceptual clarification of the notion of knowledge rather than being generated by discoveries of properties of entities belonging to the external world.

However, the fact that an important part of philosophical activity aims at producing and/or justifying overviews of certain subjects by means of conceptual clarifications does not imply that this activity can be carried out by means of entirely *a priori* procedures. In the particular case represented by a clarification of the concept of knowledge, we need to investigate the sciences and study their history. But what, in any case, remains untouched by the due attention that the epistemologist has to pay to the sciences and their history is that the object of his activity consists in producing a clarification of the concept of knowledge, a concept which we already use.

The philosophy of mathematics is no exception to this rule. In fact, 'Is mathematics conceivable as a science?', and 'How do we characterize mathematical truth?' are questions the solution of which will not generate new theorems, but produce extremely useful conceptual clarifications which will contribute to the understanding of what mathematical activity is. Of course, what has been said so far of the nature of the philosophical investigation of mathematics does not imply that the clarifications and understanding produced by it cannot and should not affect the way of doing mathematics.

If we study the history of mathematics, we realize that the different accounts of mathematical knowledge and truth offered, for example, by Platonists and Constructivists have greatly contributed to shape programmes concerning the foundations of mathematics such as Logicism and Intuitionism, which have then had a deep influence on mathematical practice. The most obvious influence of Logicism on mathematical practice is the introduction of predicate logic as a new branch of mathematics at the hands of Frege, and the other is the very pervasive use of logical notation in mathematics. In the case of Intuitionism its influence on mathematical practice has been exercized in a variety of ways through: (1) the intuitionistic development of number theory and analysis; (2) the interest that even the classical mathematician has in proving results about intuitionistic systems (completeness of predicate logic); (3) a new emphasis on rigour and on constructive procedures which has, among other things, motivated the development of classical computability theory.

Moreover, the way these programmes have fared has had important consequences on the plausibility of the philosophical conceptions of mathematical knowledge and truth which inspired them and, therefore, on the plausibility of the philosophies of mathematics of which such conceptions are part. This fact, in particular, shows that the interaction between philosophy of mathematics and mathematics works both ways.

3 The centrality of the question about realism

If one of the main tasks of the philosophy of mathematics is the production and justification of overviews of mathematics, it follows that the problems which are crucial to the formation and justification of overviews of mathematics are bound to be among the most important within the philosophy of mathematics.

But when we turn to the study of some such problems, we realize that they point at a deeper metaphysical question which is about the existence of a reality described by mathematical theories.

Indeed, if it is correct to say that mathematics is conceivable as a science then, whatever the correct definition of 'science' is, mathematical theorems have to be understood as what contributes to knowledge rather than simply being the outcome of the creative act of mathematicians. And since mathematical knowledge is expressed by mathematical statementss and mathematical statements are statements about, for instance, the existence and properties of numbers, algebraic and topological structures, sets, etc. it follows that if mathematics is a science then mathematics produces knowledge about the existence and/or properties of numbers, algebraic and topological structures, sets, etc.

However, if mathematics is not conceivable as a science, and mathematical theories are nothing but formal games or sets of constructions, mathematical theorems are nothing but the outcome of human creative activity, which is something made possible and, at the same time, limited exclusively by convention. Consequently, if mathematics is not conceivable as a science, there is no need to refer to reality to give a satisfactory account of it.

Moreover, if we believe that the truth of a mathematical statement P transcends the possibility of verifying P, it seems as if we ought to assume the existence of a reality that P is about.

On the other hand, if we believe with the verificationist that, in the absence of both a procedure of verification for it and of a refutation, a mathematical statement P lacks a truth-value, we are certainly dispensed from a commitment to the belief in the existence of a reality that P is about.

However, having mentioned the verificationist's standpoint I now need to distinguish between the verificationists who are full-blooded anti-realists and those who believe in the existence of a mathematical reality, but think that this has gaps. For the latter kind of verificationists reality is like a book whose unopened pages are blank, but as soon as the reader turns them for the first time they fill in with sentences and words which are not the outcome of arbitrary choice. (I owe this metaphor to M. Dummett.) In other words, according to the latter type of verificationists, mathematical reality is not dismissed altogether as an idle myth, but is not complete either in the sense that it might not exist a matter of fact corresponding to every meaningful mathematical question.

In relation to this second form of verificationism it is interesting to notice that a similar kind of incompleteness phenomenon may be found in the reality investigated by the empirical sciences. In fact, if the Copenhagen interpretation of quantum mechanics is correct, before a quantum measurement takes place, there is no matter of fact concerning what the outcome of the measurement is going to be.

In what I have said so far, we have seen that the two questions whether mathematics is conceivable as a science, and how mathematical truth is to be characterized, are connected with that concerning the existence of a reality described by mathematical theories. It is such a connection that shows in very clear terms the centrality of the realism/anti-realism dispute to the philosophy of mathematics. But what do we have exactly to understand by 'relism/anti-realism dispute'?

In general, we have a realism/anti-realism dispute when we try to assess either the existence of certain entities or the reducibility of these entities to others. If we accept the definition above, it follows that the term 'realism' is applicable to a wide variety of positions which arise in various branches of philosophy.

Traditionally there have been disputes between realists and nominalists over whether predicates, or universals, refer to qualities possessed by all the objects which satisfy them. If we take, for example, the word 'red', we have that, for someone who is a realist about universals, all the things of which we correctly say that they are red share a property to which the word 'red' refers. Whereas, for someone who is a nominalist about universals the term 'red' is just a word — a name — which is used to put in the same class objects which are relevantly similar to one another, but which do not actually share *one* colour-property.

A dispute about realism can arise also in science. Scientific theories often postulate the existence of entities which are beyond what we can experience with our senses. In fact, the only evidence we have about the existence of such entities is indirect, that is, it is obtained through the watching of photographs taken in bubble-chambers, or through the workings of extremely complex machines. This situation might lead some sceptic to doubt the existence of the entities postulated by scientific theories, and interpret, instead, the rôle of terms such as 'electron', 'proton', etc. as conceptual devices which can be exploited for the construction of powerful computational techniques of prediction.

The sceptic's position is greatly strengthened when it does not seem to be a trivial matter to reach an agreement among the practitioners about which,

4. THE PROBLEM OF THE FRAMEWORK 7

out of the several suggested, is the correct interpretation of the formalism of a given scientific theory. This is the present predicament of quantum mechanics.

4 The problem of the framework

The realism/anti-realism debate in the philosophy of mathematics has been raging for a long time. And the fact that it is still unresolved, in spite of the great amount of effort and ingenuity which have gone into it, has led some thinkers to believe that at the root of this unsatisfactory state of affairs lie problems of methodology.

In what follows in this chapter I shall examine the suggestions on methodology made by Carnap and Dummett, suggestions which have been proved to be among the most influential in the literature concerning the realism/anti-realism debate.

For Carnap, when we study the realism/anti-realism debate formulated in any branch of philosophy, we ought to be careful about distinguishing between a sound and an unsound form of asking questions of existence.

According to Carnap:[1]

> If someone wishes to speak in his language about a new kind of entities, he has to introduce a system of new ways of speaking, subject to new rules; we shall call this procedure the construction of a linguistic *framework* for the new entities in question. And now we must distinguish two kinds of questions of existence: first, questions of the existence of certain entities of the new kind *within the framework*; we call them internal questions; and second, questions concerning the existence or reality *of the system of entities as a whole*, called *external questions*. Internal questions and possible answers to them are formulated with the help of the new forms of expressions. The answers may be found either by purely logical methods or by empirical methods, depending upon whether the framework is a logical or a factual one. An external question is of a problematic character which is in need of closer examination.

Moreover, Carnap goes on, when it comes to the special case represented by the philosophy of mathematics:[2]

> ...nobody who meant the question "Are there numbers?" in the internal sense would either assert or even seriously consider a negative answer. This makes it plausible to assume that those philosophers who treat the question of the existence of numbers as a serious philosophical problem and offer lengthy arguments on either side do not have in mind the internal question. And, indeed, if we were to ask them: "Do you mean the question as to whether the framework of numbers, *if* we were to accept it, would be found to be empty or not?", they would probably reply: "Not at all; we mean a question *prior* to the acceptance of the new framework". They might try to explain what they mean by saying that it is a question of the ontological status of numbers; the question whether or not numbers have a certain

---

[1] See [Carnap, 1985], p. 242.
[2] See [Carnap, 1985], p. 245.

metaphysical characteristic called reality (but a kind of ideal reality, different from the material reality of the thing world) or subsistence or status of "independent entities". Unfortunately, these philosophers have so far not given a formulation of their question in terms of the common scientific language. Therefore our judgement must be that they have not succeeded in giving to the external question and to the possible answers any cognitive content. Unless and until they supply a clear cognitive interpretation, we are justified in our suspicion that their question is a pseudo-question, that is, one disguised in the form of a theoretical question while in fact it is non-theoretical; in the present case it is the practical problem whether or not to incorporate into the language the new linguistic forms which constitute the framework of numbers.

In the particular case offered by mathematics, the key notion used by Carnap to draw, in pure neo-positivistic style, a distinction between meaningful and meaningless questions of existence is that of the framework of numbers. For Carnap, a framework for the system of natural numbers[3]

> ...is constructed by introducing into the language new expressions with suitable rules: (1) numerals like "five" and sentence forms like "there are five books on the table"; (2) the general term "number" for the new entities, and sentence forms like "five is a number"; (3) expressions for properties of numbers (e.g. "odd", "prime"), relations (e.g., "greater than"), and functions (e.g., "plus"), and sentence forms like "two plus three is five"; (4) numerical variables ("m", "n", etc.) and quantifiers for universal sentences ("for every n, ...") and existential sentences ("there is an n such that ...") with the customary deductive rules.

But what is the philosophical relevance of the concept of framework and of the distinction between internal and external questions of existence?

The first thing that we must notice concerning Carnap's idea of a framework is the holism that is implicit within it. Mathematical questions of existence come, for Carnap, with the whole framework and its posits,[4] or not at all.

Secondly, the truth of a mathematical statement of existence does not transcend the use of the statement within the framework chosen.

Indeed, since a framework introduces a criterion of meaningfulness for mathematical questions in general, including questions of existence, it follows that external questions of existence are meaningless, and, therefore, the answers to them are neither true nor false.

Thirdly, there exists a strong similarity between Carnap's and Putnam's views on realism, because the idea that the truth of a mathematical statement of existence does not transcend the use of the statement within the chosen framework is at the heart of what Putnam calls 'internal realism'.[5]

---

[3][Carnap, 1985], p. 244.
[4]The posits are represented by the type of variables adopted within the framework.
[5][Putnam, 1992], p. 115:

> ...the suggestion which constitutes the essence of 'internal realism' is that truth does not transcend use.

5. THE MYTH OF THE FRAMEWORK

The confirmation of this comes from the fact that, when the framework is non-mathematical, Carnap is an internal realist in Putnam's sense as it transpires from the following quotations:[6]

> The concept of reality occurring in these internal questions is an empirical, scientific, non-metaphysical concept. To recognize something as a real thing or event means to succeed in incorporating it into the system of things at a particular space-time position so that it fits together with the other things recognized as real, according to the rules of the framework.

and[7]

> To be real in the scientific sense means to be an element of the system; hence this concept cannot be meaningfully applied to the system itself.

On the other hand, Carnap's and Putnam's positions on realism are not identical, because when the framework in question is of a mathematical nature there is, for Carnap, no commitment whatsoever to realism. The reason for this being that Carnap believes that mathematical statements are analytical, that is, he believes that mathematical statements are true (or false) simply in virtue of their meaning, and therefore have nothing to do with matters of fact.

5 The myth of the framework

Carnap's ideas about how the realism/anti-realism debate should be conducted, although very suggestive, are subject to a number of objections which undermine them among which the following.

If it is impossible to say that mathematical statements are true (or false) simply in virtue of their meaning, it follows that they have to be true (or false) in virtue of something else. This something else, which I might call 'matter of fact', would imply an ontological commitment to what is represented by the variables adopted within the framework on the part of the person who accepts the framework.

Therefore, if Carnap's view of mathematics has to be successful, in terms of being disengaged from an involvement in the traditional metaphysical dispute concerning mathematical statements of existence, mathematical statements have to be analytical. But, in the light of Quine's results on the analytic/synthetic distinction, it is more than dubious that such a thing can be shown above all if we take into account Carnap's weak reply to Quine's arguments.

For to Quine's well known claim that[8]

For a discussion of Putnam's views on realism see §5.6.
[6][Carnap, 1985], p. 243.
[7][Carnap, 1985], ibid.
[8][Quine, 1963], §IX, pp. 403-404.

we at present lack any tenable general suggestion, either rough and practical or remotely theoretical, as to what it is to be an analytic sentence ... [because] Wherever there has been a semblance of a general criterion, to my knowledge, either there has been some drastic failure such as tended to admit all or no sentences as analytic, or there has been a circularity ... or there has been a dependence on terms like 'meaning', 'possible', 'conceivable', and the like, which are at least as mysterious (and in the same way) as what we want to define

Carnap essentially replies[9]

That a certain sentence S is analytic in L_n means only something about the status of S within the language L_n; as has often been said, it means that the truth of S in L_n is based on the meanings in L_n of the terms occurring in S.

This, of course, either presupposes the existence of a matter of fact about the meanings of the terms occurring in S reative to L_n, or, if it does not, it assumes the existence of empirical criteria of synonymy for S, for terms occurring in S, etc. It is clear how both these assumptions are unwarranted in Quine's eyes.

However, a Carnapian philosopher prepared to take Quine's argument very seriously could still reply to the objection above using a typical internal realist strategy. He could, in other words, accept that, given Quine's results about the analytic/synthetic distinction, it does not make sense to say that mathematical statements are analytical, and that, consequently, one is bound, even in the mathematical case, to have an ontological commitment to believing in the entities that the framework-variables range over. 'But', he would then add, 'since also in mathematics there are cases in which, given a framework there is another competing framework which is alternative to it, and incommensurable[10] with it, we find ourselves yet again in the position of having to deny that externalist questions of existence have meaning.'

Now since the most important part of the Carnapian (internal realist) argument given above seems to be that it is impossible to have a rational discussion between incommensurable frameworks, we can attack the argument on this point showing that it must be[11]

... mistaken. For behind [the claim that it is impossible to have a rational discussion between incommensurable frameworks] there is the tacit assumption that a rational discussion must have the character of a justification, or of a proof, or of a demonstration, or of a logical derivation from admitted premises. But the kind of discussion which is going on in the natural sciences might have taught our philosophers that there is also another kind of rational discussion: a critical discussion which does not

---

[9][Carnap, 1963], p. 921.
[10]A framework F_1 is *incommensurable* with a framework F_2 when it is impossible to determine whether or not F_1 is better than F_2 simply through an appeal to the notions of meaning and derivability. For an example of incommensurable frameworks see §1.5.
[11][Popper, 1994], Chapter 2, p. 60.

seek to prove or to justify or to establish a theory, least of all by deriving it from some higher premises, but which tries to test the theory under discussion by finding out whether its *logical consequences* are all acceptable, or whether it has, perhaps, some undesirable consequences.

We thus can logically distinguish between *a mistaken method of criticizing* and *a correct method of criticizing*. The *mistaken method* starts from the question: How can we establish or justify our thesis or our theory? It thereby leads either to dogmatism, or to an infinite regress, or to the relativistic doctrine of rationally incommensurable frameworks. By contrast, the *correct method* of critical discussion starts from the question: What are the *consequences* of our thesis or our theory? Are they all acceptable to us?

A typical example of this type of rational discussion aimed at comparing two opposing frameworks within mathematics is the debate about the existence of infinitesimals which I will briefly survey in the following section.

6 A case-history: the infinitesimals

When, given a curve represented by a function f, such that $f : [a, b] \mapsto \mathbb{R}$ and $[a, b] \subseteq \mathbb{R}$, we want to determine the tangent to the curve at a point P, or when we intend to calculate the area of the portion of $\mathbb{R}^2$ enclosed by (1) the graph of f, (2) the lines drawn from the graph of f which are perpendicular to the x-axis and contain the points a and b, and (3) by the x-axis, then, given the satisfaction of certain conditions by f, the infinitary operations of differentiation and integration produce the right results.[12]

But, how are we going to understand the meaning of such operations? Infinitesimals within analysis provide a meaning, based on geometrical intuition, to the two infinitary operations:

$$\frac{dy}{dx} \quad \text{and} \quad \int_a^b f(x)dx.$$

In fact, if we postulate the existence of the infinitesimals (and of their inverses), we can interpret

$$\frac{dy}{dx}$$

as the *value* of the ratio

$$\frac{f(x + dx) - f(x)}{dx}$$

---

[12]What I mean by this is that the results we obtain in geometry are also obtained in analysis and, moreover, when we apply the techniques of analysis to practical problems, e.g., in calculating the velocity of a falling body or the area of a field or the volume of a wine cask, the *predictions* we make by means of our calculations are then experimentally confirmed.

when dx is *infinitesimally small*; and

$$\int_a^b f(x)dx$$

as the sum of an infinite number of rectangles having height $f(x)$ and width dx, for dx *infinitesimally small*.

Let us now see how infinitesimals are used in the solution of the problem of calculating the tangent to a point P of a parabola. This will be useful to enable us to focus on some of the points raised in the controversy about the existence of infinitesimals.

If we represent the branch of the parabola $x^2 = y$, for $x \geq 0$, in Cartesian coordinates, and pick a point P on the parabola, Newton provides an algorithm to calculate the tangent $\mathcal{T}$ to the parabola at P (see fig. 1.1).

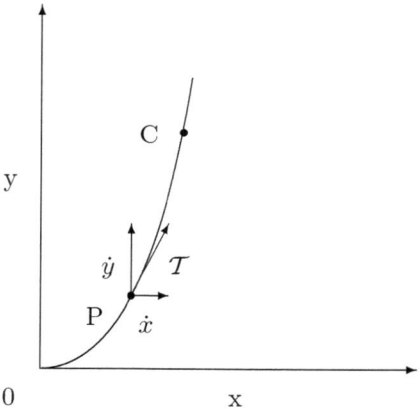

Figure 1.1. Parabola

If we think of the branch of parabola in $\mathbb{R}^2$ as the motion of a point C, such a motion can be expressed as the composition of the motions of two points A and B such that: (1) A moves horizontally for a distance x and B moves vertically for a distance $y = x^2$ in equal intervals of time; (2) A has a velocity vector with length $\dot{x}$ (Newton's fluxion), whereas B has a velocity vector with length $\dot{y}$. In modern notation

$$\dot{x} = \frac{dx}{dt} \text{ and } \dot{y} = \frac{dy}{dt},$$

where dt is an interval of time.

From the considerations above and the application of the parallelogram rule to the velocity vectors, it follows that: (i) the tangent $\mathcal{T}$ to the parabola

6. A CASE-HISTORY: THE INFINITESIMALS

at P is the velocity vector which is the sum of the horizontal and vertical vectors whose lengths are respectively $\dot{x}$ and $\dot{y}$ when dt is infinitesimally small, that is, $dt = o$; (ii) the slope of the tangent is, therefore, $\dot{y}/\dot{x}$, where

$$\frac{\dot{y}}{\dot{x}} = \frac{dy}{dx}. \tag{1.1}$$

Hence, to solve our problem we have to compute $\dot{y}/\dot{x}$. This is done in the following way:

$$y = x^2, \tag{1.2}$$
$$y + \dot{y}o = (x + \dot{x}o)^2, \tag{1.3}$$
$$y + \dot{y}o = x^2 + 2x\dot{x}o + \dot{x}\dot{x}oo, \tag{1.4}$$
$$\dot{y}o = 2x\dot{x}o + \dot{x}\dot{x}oo, \tag{1.5}$$
$$\frac{\dot{y}}{\dot{x}} = 2x + \dot{x}o, \tag{1.6}$$
$$\frac{\dot{y}}{\dot{x}} = 2x. \tag{1.7}$$

(1.2) is the equation of the parabola; (1.3) and (1.4) are trivial; (1.5) is obtained from (1.4) and (1.2); (1.6) derives from dividing (1.5) by $\dot{x}o$; (1.7) is justified by the fact that we can cancel out $\dot{x}o$ in (1.6) because it is infinitesimally small.

At this point we must observe that the method illustrated in the example above: (1) provides a procedure for calculating the tangent and not simply a way of showing its existence and uniqueness; (2) introduces a way of calculating which involves the use of infinitesimal quantities such as $\dot{x}o$ and $\dot{y}o$.

As Giorello points out, infinitesimals play in the analysis of Newton and Leibniz a rôle similar to that performed by Greek atomism in physics:[13]

> No less than ancient atomism ... the appeal to mathematical indivisibles and infinitesimals in dealing with geometrical and kinematical problems was indeed building up an unknown and invisible world behind the world that is known to us.

However, in the course of time several philosophers, among whom was prominent Bishop Berkeley, and mathematicians, raised objections against the acceptability of the concept of infinitesimals. Although, as we shall see in what follows, such objections were formulated mainly by people who did not share the new Newtonian-Leibnizian framework, they were far from being meaningless, or unfruitful, and gave rise to a rational debate, which very much affected the developement of mathematical analysis.

---
[13][Giorello, 1995], §8.3.1, p. 141.

14   THE REALISM/ANTI-REALISM DEBATE

As is well known, the fire of controversy over infinitesimals first raged within the French Academy of Sciences, and saw the faction of the 'finitists' oppose that of the 'infinitesimalists', as Mancosu calls them.[14] The finitists were those mathematicians who were anchored to the old 'Cartesian refusal to admit infinitary mathematics as a rigorous discipline ... ';[15] whereas the infinitesimalists were those mathematicians who had embraced Leibniz's calculus. The dispute between finitists and infinitesimalists was over the existence of infinitesimals and the mathematical rigour of the algorithms involving them, and had at its heart a metaphysical issue about infinitesimals as existing objects. This is very clearly shown by the attitude of the Marquis de L'Hôpital who, when Leibniz asserted that it is possible to give a fictionalist account of his mathematical analysis, which does not base it on metaphysical assumptions,[16] thought that Leibniz was betraying the cause of the new calculus.

Metaphysical issues are also at the very centre of Berkeley's *Analyst* where the empiricist philosopher argues, among other things, that:[17]

> ... he who can digest a second or third fluxion, a second or third difference, need not, methinks, be squeamish about any point in divinity.

One of the standard complaints of the finitists about infinitesimals was that their very definition given in terms of *vanishing increments* seemed to be inconsistent. In fact, they reasoned, if an increment vanishes, how can we possibly call it an 'increment'?

Another important objection against the *modus operandi* with infinitesimals was that in transformations such as the one of p. 13, the quantity $\dot{x}o$ is considered in (1.5) as different from 0 (because we obtain (1.6) dividing

---

[14][Mancosu, 1996], p. 165.

From 1700 to 1706 the academy was divided over the admissibility of the new techniques: on one side stood the infinitesimalist group characterized by its total adherence to the new Leibnizian calculus in the version codified by L'Hôpital and in general by a commitment to the existence of infinitesimal quantities; on the other side, the finitist faction characterized by a refusal to give a rigorous status to infinitesimal considerations and by a general adherence to classical techniques.

[15][Mancosu, 1996], p. 177.
[16][Mancosu, 1996], p. 172:

(a) There is no need to base mathematical analysis on metaphysical assumptions.
(b) We can nonetheless admit infinitesimal quantities, if not as real, as well-founded fictitious entities, as one does in algebra with square roots of negative numbers ... Or (c) one could organize the proofs so that the error will be always less than any assigned error.

[17][Berkeley, 1734], §7, p. 65.

6. A CASE-HISTORY: THE INFINITESIMALS

both sides of (1.5) by $\dot{x}o$), and in (1.6) as equal to 0, leading once again to inconsistency in the understanding of $\dot{x}o$.[18]

These and other critical remarks, which were part of a long lasting rational debate, led the mathematical community to wonder whether the pay off of infinitesimals, in terms of making sense in intuitive/geometrical terms of the fundamental analytical operations, and of providing an effective heuristic system for the discovery of new results, could really counterbalance the problems generated by the clear lack of logical plausibility.

All the worries of those analysts who had logical rigour at heart were eventually put to rest by the development of the theory of limits, and by its application to analysis. Such a theory, in fact, provides a definition of differentiation and integration which is independent of the infinitesimals, showing, by a most emphatic application of Ockham's razor, that the notion of infinitesimal is redundant within analysis.

In fact, if we interpret dy/dx and $\int_a^b f(x)dx$ as:

$$\frac{dy}{dx} = \lim_{\triangle x \to 0} \frac{\triangle y}{\triangle x} = \lim_{h \to 0} \frac{f(x+h) - f(x)}{h} = f'(x),$$

and

$$\int_a^b f(x)dx = \lim_{n \to \infty} \sum_{i=1}^n f(x_i) \triangle x_i;$$

we have that: the derivative is nothing but the limit of a function; and that the integral is the limit of a sequence of finite sums.

The case-history about infinitesimals is very important with regard to the tenability of Carnap's *veto* against external questions of existence, because it shows that the mathematical community engages in a rational debate about the relative merits of two opposing frameworks, a debate which develops in the direction pointed out by Popper through an attempt to answer questions such as 'What are the consequences of our theory?', 'Are they all acceptable to us?'. This is precisely what has happened in the case of the infinitesimals and shows that asking external questions of existence, which is equivalent to questioning a framework as a whole, is something that makes mathematical sense, is a very important part of mathematical activity, and is dealt with by the mathematical community through the use of rational criteria.

To such an interpretation of the case-history discussed in this section someone may object that, since the theory of infinitesimals has been resurrected from its ashes by A. Robinson and others, the two above mentioned opposing frameworks for analysis turn out to be incommensurable after all.

---

[18] See on these and other objections to the meaningfulness of operating with infinitesimals [Boyer, 1949], Ch. VI, pp. 224–266.

In fact, if frameworks for analysis were truly commensurable they would have to give origin to a linearly ordered sequence (of frameworks)

$$F_1, F_2, \ldots, F_n$$

such that F_{i+1}, where $1 \leq i < n$, is *better* than F_i. But this is plainly not the case, because the theory of infinitesimals which at a certain point in time seemed to have been discarded in favour of the theory of limits has made its come back at a later stage.

However, the reason why this objection does not work is that the postulation of infinitesimals does not in itself constitute a framework for analysis. On the other hand, we can rightly call 'framework' Newton's and Leibniz's systems of analysis based on the theory of infinitesimals and Weierstrass's system of analysis based on the theory of limits, because we can develop analysis within them. And if we do this, there is no doubt that the mathematical community compares frameworks with one another in relation to their consequences and judges the Weierstrassian framework for analysis to be better than Newton's and Leibniz's.

Moreover, it is important to observe that Robinson's framework for analysis is very different from Newton's and Leibniz's; I shall here mention two such differences.

First, Robinson's system (framework) of non-standard analysis is based on the proof of the existence of a non-standard model of the theory of real numbers. (Such a proof essentially depends on the compactness theorem of first-order logic.)

Secondly, the successful attempt to avoid the traditional anomalies which beset Newton's and Leibniz's systems led Robinson to define very important notions, such as that of derivative, in a very different way from that present in the systems of Leibniz and Newton. For example, whereas for Leibniz the derivative is the value of the ratio

$$\frac{dy}{dx},$$

where dx is infinitesimal, for Robinson, instead, the derivative is the *standard part* of the ratio dy/dx, where dx is infinitesimal.[19]

---

[19]If $\mathbb{R}^*$ is the domain of the non-standard model of the theory of real numbers, call

$$\mathcal{F} = \{x \in \mathbb{R}^* : |x| < y, \text{ for some } y \in \mathbb{R}\}$$

the set of *finite* elements of $\mathbb{R}^*$; and

$$\mathcal{I} = \{x \in \mathbb{R}^* : |x| < y, \text{ for all positive } y \in \mathbb{R}\}$$

the set of the *infinitesimal* elements of $\mathbb{R}^*$.

7. FROM LOGIC TO METAPHYSICS? 17

Having shown that Carnap's approach to the realism/anti-realism debate in the philosophy of mathematics is not entirely satisfactory, I shall now move on in the next section to the study of a different way of setting up the realism/anti-realism dispute in the philosophy of mathematics.

7 From logic to metaphysics?

According to Michael Dummett, if you are a metaphysical realist about mathematics, that is, if you believe in the existence of mathematical objects or structures, you must also be a realist about mathematical truth, in other words, you must believe that the truth of a mathematical statement transcends verification.

In fact, if you believe in the existence of numbers or of a number-theoretical structure then, for Dummett, you must also believe that number-theoretical statements are true or false according to whether they correctly or incorrectly describe properties of numbers or of the number-theoretical structure. From this follows that, if you are a metaphysical realist then, for you, the truth or falsity of a number-theoretical statement is independent of, transcends the availability of methods apt to determine the statement's truth-value.

The position described above has two main consequences for a metaphysical realist about number theory. First, a proof is, for him, *only* a way of justifying the assertion of a number-theoretical statement and, secondly, number-theoretical statements for which no method of decision is known are considered by him to be determinately true or false.

It is important to notice that the latter point implies the acceptance of the principle of Bivalence within number theory, i.e., the acceptance of the idea that, for every number-theoretical statement S, S is determinately true

---

If by *standard* x we mean any x such that $x \in \mathbb{R}$, since each finite x can be uniquely decomposed into

$$x = s + i,$$

where s is standard and i is infinitesimal, the function $st : \mathcal{F} \to \mathbb{R}$ maps elements of $\mathcal{F}$ into their standard parts and has the following characteristics:

$$\begin{aligned}
st(x) &= x, \quad \text{if } x \in \mathbb{R}; \\
st(x) &= 0, \quad \text{if } x \text{ is infinitesimal}; \\
st(x+y) &= st(x) + st(y); \\
st(x \times y) &= st(x) \times st(y).
\end{aligned}$$

Now, if you apply this to (1.6), p. 13, you have:

$$\begin{aligned}
st(\tfrac{\dot{y}}{\dot{x}}) &= st(2x + \dot{x}o) \\
&= st(2x) + st(\dot{x}o) \\
&= 2x + 0 \\
&= 2x.
\end{aligned}$$

or false, and that, for Dummett, the acceptance of the principle of Bivalence entails the acceptance of classical logic, because the principle of Bivalence validates the law of Excluded Middle, and other laws of classical logic.

By the same token, if you are a metaphysical anti-realist about mathematics — if you do not believe in the existence of mathematical objects and structures — then, for Dummett, you must also be an anti-realist about mathematical truth, that is, a mathematical statement is true for you just in case there is a constructive proof of it, and false just in case there is a refutation of it.

For, since according to the metaphysical anti-realist about number theory there are neither numbers nor a number-theoretical structure then, for him, there cannot be number-theoretical truth either independently of the existence of methods of decision/verification for number-theoretical statements. One of the conclusions that Dummett draws from this is that the anti-realist's notion of number-theoretical truth collapses on to that of proof.

Consequently, for a metaphysical anti-realist about number theory, a constructive number-theoretical proof is *constitutive* of the concept of number-theoretical truth and, therefore, statements like the twin-primes conjecture, the Goldbach conjecture, the Riemann Hypothesis, etc. are neither true nor false. From the latter point follows that, for a metaphysical anti-realist about number theory, the principle of Bivalence does not hold in number theory and, therefore, according to Dummett, the logic that, for the metaphysical anti-realist about number theory, regulates the language of arithmetic is non-classical.

For Dummett, the considerations above have two important consequences. First, the traditional debate between metaphysical realists and anti-realists concerning number theory can be translated into another type of debate about whether it makes sense to say that number theoretical statements for which we do not have a procedure of decision — these are the elements of the so-called 'disputed class' — are true or false.

Secondly, if you want to argue against metaphysical realism (or anti-realism) concerning number theory, you can just as well argue against the adequacy of a realist (or anti-realist) conception of truth for the number theoretical statements belonging to the disputed class and, therefore, as we have seen above, also about whether certain '...forms of deductive arguments ... are to be accepted as valid'[20] in doing number theory.

The question that at this point naturally arises is the following: how can we argue in favour of or against a realist or an anti-realist view of arithmetical truth? For Dummett, we can address this problem by means of a theory of meaning for the language of arithmetic. But, what is a theory

---

[20][Dummett, 1991], p. 16.

7. FROM LOGIC TO METAPHYSICS? 19

of meaning? And, what has meaning got to do with truth?

According to Dummett, the fundamental task of a theory of meaning is that of determining[21]

> whether, as the realist believes, our understanding [the meaning] of mathematical statements demands to be explained in terms of a grasp of what would render them true, independently of our knowledge of their truth-value, and, if so, in what our grasp of this consists; or whether, as the constructivist supposes, it can be sufficiently explained in terms of our ability to recognise proofs or disproofs of such statements when presented with them.

What I have called 'the fundamental task of a theory of meaning' shows in very clear terms the strong relationship existing between meaning and truth. Moreover, achieving it would, among other things, provide us with a compelling justification for determining the meaning of the logical constants, and this, given the systematic nature of the language of arithmetic, would '...supply a general characterization of the contribution which a sentence makes to the content of a more complex sentence formed from it.'[22]

One of the main consequences of determining the meaning of the logical constants would then be the possibility of individuating the logic regulating the language of arithmetic. In fact, as is well known, the logical laws which regulate the language of standard arithmetic, e.g., the Law of Excluded Middle, the law of Double Negation elimination, etc., depend on the 'truth-table meaning' of the logical constants used in the sentences of the language of standard arithmetic. If we were to change such a meaning according, for instance, to the Brouwer-Heyting-Kolmogorov (BHK) interpretation of the logical constants, the door would be open to the construction of the so-called 'weak counter-examples' to the law of Excluded Middle.

But how can a meaning theory hope to accomplish the fundamental task of a theory of meaning without being itself biased in a realist or an anti-realist sense concerning the notions of meaning and truth?

To answer this question Dummett takes a Wittgensteinian stand on meaning, that is, for him,[23]

> No hidden power confers...meanings on [the statements belonging to the disputed class]: they mean what they mean in virtue of the way we use them, and of nothing else

and asserts that a theory of meaning aims only at elucidating our linguistic practices.

---

[21][Dummett, 1991], p. 14.
[22][Dummett, 1991], p. 14.
[23][Dummett, 1991], p. 13.

'But', someone might say, 'if the meaning of statements is established by the way we use them, why should there be a genuine controversy about such meanings?' To this objection Dummett has a reply, and this is that[24]

> Although we know what they mean and have come, in the course of our childhood and our education, to learn what they mean, we do not know how to represent their meanings: that is, we learn to use them but do not know precisely what it is that we learn when we learn that.

Therefore, for Dummett, if we want[25]

> To gain a complete understanding, to come to command a clear view of how they function, we need to scrutinize our own linguistic practices with close attention, in order, in the first instance, to become conscious of exactly what they are, but with the eventual aim of attaining a systematic description of them. Such a description will give a representation of what it is for the words and expressions of our language to have the meanings that they do. It must embrace everything that we learn when we first learn language, and hence cannot take as already given any notion a grasp of which is possible only for a language-speaker. In this way, it will lay bare what makes something a *language*, and thus what it is for a word or sentence to have a meaning.

According to Dummett,[26]

> success [in the task of constructing a meaning-theory] is to be estimated according as the theory does or does not provide a workable account of a practice that agrees with that which we in fact observe.

8 Assessing some Dummettian claims

At this point someone may try to undermine the Dummettian realism/anti-realism debate saying that the anti-realist about truth '*must* mean something different [from the realist] by the logical constants'[27] and, therefore, he is not really casting doubts on the laws of classical logic, but developing a different system altogether, a system within which the *literal translations* of some of the laws of classical logic happen not to hold. This is, indeed, the attitude that several mathematicians, among whom is prominent D. Bridges, manifest towards the incompatibility of some of the results of classical logic and mathematics with those of intuitionistic logic and mathematics.[28]

---

[24][Dummett, 1991], p. 13.
[25][Dummett, 1991], p. 13.
[26][Dummett, 1991], pp. 13–14.
[27][Dummett, 1991], p. 17.
[28]See [Bridges, 1998], p. 55:

> For several decades after the publication of his thesis ..., Brouwer developed mathematics based on his philosophy of *Intuitionism*, a philosophy that led him to a number of concepts and principles which, on first reading, appear to contradict classical mathematics (CLASS). For example, a theorem of INT [intuitionistic mathematics] states that

8. ASSESSING SOME DUMMETTIAN CLAIMS

It is clear that, if the position I have described above, and which I am going to call 'tolerant pluralism', is correct then systems of classical and intuitionistic logic, classical and intuitionistic number theory, classical and intuitionistic analysis, etc. turn out to be incommensurable with one another. And, of course, if these systems are incommensurable with one another, and are consistent, then vain is the hope of setting up a Dummettian realism/anti-realism debate to decide which concept of truth and which logic — classical or intuitionistic — is the right one for the elements of a given disputed class of mathematical statements.

It is important to notice that, although within tolerant pluralism the traditional metaphysical question about the existence of mathematical objects and structures remains open, there is no longer a connection between the metaphysical issue of realism and the problem concerning the nature of mathematical truth.

With regard to the first point, consider that a tolerant pluralist in mathematics can be either a metaphysical realist or an anti-realist. A clear example of an anti-realist tolerant pluralist is that of the formalist. In fact, the formalist is a metaphysical anti-realist, because, for him, mathematical theories are just formal 'games', and is also a tolerant pluralist, because, according to him, any consistent 'mathematical game' is admissible. An example of a tolerant pluralist of realist inclinations is, instead, the philosopher of mathematics who adopts what M. Balaguer calls 'full-blooded Platonism'[29] (**FBP**) and P. Maddy 'plentiful Platonism', that is,[30]

> ...the view that there exists an objective world of sets corresponding to each and every consistent theory in a first-order language with $\in$ as its sole non-logical symbol.

Concerning, now, the second point, it is very instructive to examine the case of the formalist again. It is well known that, although the formalist plays some mathematical *games* using classical logic and mathematics, and others using intuitionistic logic and mathematics, this type of activity does not at all weigh on his anti-realist conscience. The reason for this is that, according to him, mathematical theories and logic are the outcome of mere convention.

(*) Every function from $[0, 1]$ to the real line $\mathbb{R}$ is uniformly continuous.

The apparent absurdity of this statement is, however, illusory, as is suggested by the following more careful re-statement of it:

(**) Every intuitionistically defined function from the intuitionistic interval $[0, 1]$ to the intuitionistic real line is, intuitionistically, uniformly continuous.

[29] See [Balaguer, 1998] and §3.10 of this book.
[30] [Maddy, 1998], §1, p. 162.

On the other hand, also the plentiful Platonist is not bothered by having to work in some mathematical theories using classical logic and mathematics, and in others using intuitionistic logic and mathematics. This apparent peculiarity in his behaviour is satisfactorily accounted for, if we remember that the plentiful Platonist believes that any consistent mathematical theory expressed in a first-order language, with '$\in$' as its sole non-logical symbol, correctly describes some features of an existing universe of sets.

From these last considerations follows that, within tolerant pluralism, we can be metaphysical realists or anti-realists independently of which conception of truth and logic we happen to be working with.

Dummett has reacted to the threat posed by tolerant pluralism to his attempt to find a pathway leading from logic to metaphysics saying that, in a *genuine* realism/anti-realism dispute about truth, the participants in the dispute deny that their opponents have 'hold of a coherent meaning'.[31]

But is it always the case that, whenever we have a metaphysical realism/anti-realism dispute, we can also have a genuine realism/anti-realism dispute about truth? This question is rather important, because if the answer to it were 'No', this would not only show the existence of a serious limitation in the range of applicability of the Dummettian programme, but would also expose the absence of a general connection between the problem of understanding the nature of mathematical truth and that regarding the existence of mathematical objects and structures.

A consequence of a result proved by Gödel (in [Gödel, 1933e]) about the possibility of translating the wff. of a first-order formal system of classical arithmetic (**PA**) into the wff. of a first-order formal system of intuitionistic arithmetic (**HA**), is that **PA** and **HA** are equi-consistent. The equi-consistency of **PA** and **HA** and, in particular, the consistency of **PA** relative to **HA**, is very important for our discussion. For it reveals that, in spite of the mutually exclusive conceptions of meaning, truth, etc. adopted in the two formal systems, **HA** validates the patterns of deductive reasoning adopted within **PA** — that is, classical logic — in the sense that, if **HA** is consistent then applying correctly classical logic within **PA** cannot lead to inconsistency.

Now, if it is the case that **HA** validates the patterns of deductive reasoning adopted within **PA**, and *vice versa*, we cannot have a genuine Dummettian style realism/anti-realism dispute about number theory. For, how can the participants in the dispute deny that their opponents have hold of a coherent meaning? Doing this would be as unreasonable as having supporters of Euclidean geometry who deny the meaningfulness of the notions involved in the development of hyperbolic geometry even after Euclidean

---

[31][Dummett, 1991], p. 17.

8. ASSESSING SOME DUMMETTIAN CLAIMS

models of hyperbolic geometry have been discovered.

On the other hand, it is, indeed, possible to adopt tolerant pluralism with regard to **PA** and **HA** engaging in a genuinely metaphysical dispute about realism, which sees the formalist oppose the plentiful Platonist.

Dummett might, now, reply that the equi-consistency of **PA** and **HA** is a result which specifically involves certain formal systems of arithmetic, and that the intuitionist believes that mathematical theories cannot be entirely captured within formal systems.

This counter-objection can be met, though, if we observe that the equi-consistency of **PA** and **HA** is a hard mathematical fact that clearly speaks of a probable equi-consistency of classical and intuitionistic number theory, a hard mathematical fact that cannot be quickly brushed aside without providing mathematical evidence in favour of the idea that classical number theory and intuitionistic number theory are not equi-consistent after all.

A possible attempt to by-pass the above mentioned difficulties might start from the assumption that meaning is use, and the concession that we can have a dispute about arithmetical truth, even though the disputants do *not* deny the coherence of their opponent's conception of arithmetical truth.

A dispute about arithmetical truth based on the above assumption and concession would, indeed, be sound, if the aim of the disputants were no longer that of determining 'the' nature of arithmetical truth, and 'the' meaning of arithmetical statements, but the more modest one of individuating the account of arithmetical truth, and meaning, which agrees best with arithmetical practice.

Taking for granted that what I have just sketched above works as a way out from the objections and exceptions I moved to the soundness and generality of the Dummettian realism/anti-realism dispute about truth in mathematics, it is, now, legitimate to wonder whether making use of such a way out is going to allow us to describe a pathway leading from logic to metaphysics.

One of the problems connected with it is that, although Dummett and Wittgenstein and others talk about arithmetical practice, it is not clear what they mean by it. Is arithmetical practice what we do when we learn to add, multiply, etc. or is it, rather, what a mathematician does when he contributes to number theory?

If it is the former, then we might indeed succeed in extricating from the study of the psycho-socio-anthropo-logico-syntactic-etc. tangle of factors involved in mastering arithmetical practice a precise idea about which concept of truth and meaning assigned to the statements belonging to the disputed class of arithmetical statements would fit best the arithmetical practice of the day. But would this help in any way to sort out the metaphysical question about whether or not numbers exist?

It becomes clear that it would not, as soon as we realize that, in addressing the metaphysical question of realism, we are interested in producing information concerning the existence of certain entities: numbers, structures, etc.; and that this is information which, as Kant famously said in his criticism of the ontological argument, we cannot find in concepts, and therefore, we cannot find, in particular, in the concepts of number, structure, and, we may add, arithmetical practice.

For:[32]

> By whatever and by however many predicates we may think a thing — even if we completely determine it — we do not make the least addition to the thing when we further declare that this thing *is*.

From this follows that the only thing a consistent set of number theoretical concepts, taken together with concepts of truth and meaning which snugly fit the arithmetical practice of the day, can offer is a representation of how things *could* be, if there were (or if there were not) numbers, structures, etc.

If, on the other hand, by 'arithmetical practice' we mean what the mathematician does when he contributes to number theory then, again, reflecting on the mathematician's training, cultural milieu, etc. might help us at the very most to understand why he has, for instance, become a Bourbakist, but, for the same reasons I gave above, it would not help us to decide whether mathematics is a science of structures.

A different argument against the soundness of the Dummettian debate goes as follows.[33]

Let us consider a Dummettian realism/anti-realism dispute in any branch of mathematics, and assume that the anti-realist disputant is an intuitionist. As a consequence of the intuitionistic meaning of negation, for the intuitionist, there cannot be absolutely undecidable mathematical statements, that is, mathematical statements which neither can be known to be true nor can be known to be false.

How does the intuitionist interpret this result? An obvious reply is that, for the intuitionist, the result above speaks of the open texture and power of the creative process at the root of mathematical activity which is such that sometime in the future, someone, somewhere might invent new constructive proof-procedures which will enable him or others to prove or refute the statements belonging to the disputed class.

---

[32][Kant, 1787], Transcendental Dialectic, Book II, Chapter III, §4, p. 505.

[33]This argument has been produced on the basis of contributions given in [Wright, 1993a], [Wright, 1993b], [Rasmussen & Ravnkilde, 1982], [Martin-Löf, 1994], and in [Prawitz, 1998].

8. ASSESSING SOME DUMMETTIAN CLAIMS 25

However, such an answer is unsatisfactory, because, among other things, it appears to have the unacceptable consequence of introducing tense into intuitionistic mathematics.

To see how unacceptable this is for the intuitionist, consider that when the intuitionist conjectures that there are infinitely many twin-primes[34]

> ... he is ... not making the conjecture that *it will be proved that there are infinitely many twin-primes* [the italics are mine], which is a conjecture about future history ... [But that] it is provable that there are infinitely many twin-primes.

Where can the intuitionist go from here without either becoming a victim of the objection above or falling into outright realism? If the intuitionist accepted the assumption that proofs and refutations exist before we find them, such an assumption would guarantee that, for each statement of the disputed class, there would be a proof or a refutation before we find it; and, it would also avoid introducing tense into intuitionistic mathematics, because, since proofs exist before they are found, a statement belonging to the disputed class was, is and always will be either provable or refutable.

An indirect confirmation of the importance, for the intuitionist, of the assumption concerning the existence of the realm of proofs is that this provides the solution of another serious problem for the intuitionist, that of the objectivity of mathematical proofs.[35]

Indeed, as Prawitz says:[36]

> ... something is a proof in mathematics if it is either a canonical proof or a method for finding a canonical proof. We may claim that once we have laid down what counts as canonical proofs, it is a factual matter whether an alleged proof amounts to such a canonical proof. If it is not a canonical proof, then it is again a factual matter whether the alleged proof yields a method for finding a canonical proof. *Hence it should be clear that it is not our treating it as a proof that makes it a proof.* [The italics are mine] This seems a reasonable claim. It makes something a proof in virtue of the meaning of the expressions involved, which is also reasonable. But it also seems to imply that the question of whether something is a proof is fixed when the meanings are given, that is, when it is given what counts as a canonical proof. From this it is natural to conclude that already, before a proof of a sentence is found, it is determined that there is such a proof. Provability, which I want to identify with truth, becomes in this way something objective.
>
> My point is that the same features which make proof and to be proved objective notions also make provability objective. If we discard the notion of provability, maintaining that before a sentence is proved it is not determined whether it is provable, then it seems that we pull away the grounds for the objectivity of proofs.

---

[34][Prawitz, 1998], §. 4, p. 47.

[35]That the objectivity of mathematical proof is a serious problem for any constructivist and, therefore, in particular for the intuitionist, is shown by the fierce controversy raging between intuitionists, followers of Markov's school of constructive mathematics, and strict finitists over what must we mean by 'mathematical construction'. See on this §§4.8–4.12.

[36][Prawitz, 1998], §. 5, p. 49–50.

Now, if it is the case that believing that proofs exist even before we find them is not sufficient to persuade the intuitionist to accept what *he* understands as the law of Excluded Middle, it is certainly the case that postulating a realm of proofs, the elements of which exist before being known, is plainly incompatible with the intuitionist's anti-realist view of mathematics, in spite of Prawitz's reassuring remark that 'proofs as here understood are something that in principle can be known by us'.[37] Is this incoherence unavoidable within intuitionist philosophy of mathematics? My claim is that it is not.

To see this consider that, for Brouwer, the elements of a Dummettian disputed class of mathematical statements have no mathematical content whatsoever.[38] Therefore, according to Brouwer, it is meaningless to ask whether they are provable or not and, *a fortiori*, whether the principle of Bivalence applies to them or not, etc.

From this we can conclude that the incoherence highlighted above does not affect Brouwer's Intuitionism, but only Dummett's. And the reason for the emergence of such an incoherence within Dummett's system is precisely the main distinguishing factor between his version of Intuitionism and Brouwer's: language. Indeed, as Prawitz says, Dummett's Intuitionism is[39]

> ... based on considerations of meaning approached from a verificationist point of view rather than on considerations of an ontological kind.

The main consequence of what has been argued above seems to be that, independently of whether or not there is a pathway leading from logic to metaphysics, it appears that Michael Dummett has not yet succeeded in showing there is one. But, perhaps, it is possible to produce a strengthening of this conclusion.

9 A case in favour of independence

Indeed, if we represent the possible combinations of being metaphysical realists (**MR**) and anti-realists (**MA**), with being realist (RT) or antirealist about truth (AT) as in fig. 1.2

---

[37][Prawitz, 1998], §. 4, p. 48.

[38]In [Brouwer, 1981], Lecture 1, p. 6:
> ... let us pass to infinite systems and ask for instance if there exists a natural number n such that in the decimal expansion of π the nth, $(n + 1)$th, ..., $(n + 8)$th, and $(n+9)$th, digits form the sequence 0123456789. This question, relating as it does to a so far not judgeable assertion, can be answered neither affirmatively nor negatively. But then, from the intuitionist point of view, because outside human thought there *are* no mathematical truths, the assertion that in the decimal expansion of π a sequence 0123456789 either does or does not occur is *devoid of sense* [The italics are mine].

[39][Prawitz, 1998], p. 41.

9. A CASE IN FAVOUR OF INDEPENDENCE

$\Re$	RT	AT
MR	p_{11}	p_{12}
MA	p_{21}	p_{22}

Figure 1.2.

we realize that in philosophy of mathematics all four possibilities present in the matrix above are realized by positions which, in so far as the particular combination adopted of **MR/MA** with RT/AT is concerned, are perfectly coherent.

To see this, first of all, consider that the belief that mathematics is a science of abstract objects and the belief that mathematical truth transcends verification (possibility p_{11}) can both be true in a situation in which numbers, structures, etc. exist independently of mathematical activity. In such a situation, in fact, it is possible that, for instance, the belief that there are infinitely many pairs of twin-primes[40] is either true or false independently of our ability to verify (refute) it (see §1.7).

A position in the literature which upholds possibility p_{11} is mathematical Platonism. According to the Platonist, numbers, structures, etc. are abstract entities which exist independently of being represented, and the truth of mathematical statements transcends verification. This latter feature of mathematical Platonism clearly emerges, for instance, within Gödel's Platonist philosophy of mathematics, as the well known view that the Continuum Hypothesis is conceivably either true or false, even though it is independent from the Zermelo-Fraenkel axioms.[41]

Secondly, a case which reveals that metaphysical anti-realism and realism about truth (possibility p_{21}) are compossible is that of empirical formalism. This is a position defended in the literature by H. Curry.

The empirical formalist believes that mathematics is a *science of formal systems*, a science which includes among its statements some which possess

---

[40] A pair of *twin-primes* is a pair of successive odd integers p and $p + 2$ such that both p and $p + 2$ are primes.

[41] [Gödel, 1964], pp. 266-7:

...it has been suggested that, in case Cantor's continuum problem should turn out to be undecidable from the accepted axioms of set theory, the question of its truth would lose its meaning, exactly as the question of the truth of Euclid's fifth postulate by the proof of the consistency of non-Euclidean geometry became meaningless for the mathematician. I therefore would like to point out that the situation in set theory is very different from that in geometry, both from the mathematical and from the epistemological point of view.

At some point, Gödel thought that the true power of the continuum is $\aleph_2$. See [Gödel, 1970].

a meta-theoretic nature. Such a view justifies us in saying that the empirical formalist is both a metaphysical anti-realist and a realist about truth.

For, on the one hand, the empirical formalist believes that, for example, first-order formal systems of arithmetic are games whose rules are entirely based on convention. And, on the other hand, his view that 'mathematics ... is a body of propositions dealing with a certain subject matter; ... propositions [which] are true in so far as they correspond with the facts',[42] clearly expresses a verification transcendent conception of truth.

It is very important to notice here that the above mentioned realist conception of mathematical truth is perfectly consistent with metaphysical anti-realism, because the facts described by true mathematical statements are not made of numbers, structures, etc. but consist of elements of formal systems.

Thirdly, an example of a standpoint which reconciles metaphysical realism with anti-realism about truth (possibility p_{12}) is that of the philosopher who believes in the existence of the natural numbers, but, at the same time, holds that the truth of a statement about natural numbers does not transcend the statement's conditions of verifiability.

Such a person might justify his metaphysical realism about natural numbers by means of a transcendental argument of the following kind:[43]

> The positive integers and their arithmetic are presupposed by the very nature of our intelligence and, we are tempted to believe, by the very nature of intelligence in general.

And then explain, convincingly, and consistently with his metaphysical realism, his anti-realism about truth arguing that when we consider mathematical statements, we realize that some of them[44]

> ... are merely evocative, ... [that is, they] make assertions without empirical validity. [But that] There are also mathematical statements of immediate empirical validity, which say that certain performable operations will produce certain observable results: for instance, the theorem that every positive integer is the sum of four squares.

Now, since mathematics proper is made of statements which have immediate empirical validity — they are verifiable, i.e., they are provable constructively or are refutable — it follows that it makes sense to attribute truth-values only to statements such as these.

Consequently, as Hilbert said of ideal statements, mathematical statements which are merely evocative are neither true nor false, their importance

---

[42][Curry, 1954], p. 202.
[43][Bishop & Bridges, 1985], p. 4.
[44][Bishop & Bridges, 1985], p. 2.

simply consisting in making mathematicians' life easier through simplifications. (The positions in the literature which come closest to that described above are Kronecker's and Bishop's.)

Lastly, I shall examine the case of a philosopher of mathematics who is a metaphysical anti-realist, as well as being an anti-realist about truth: L. E. J. Brouwer.

Brouwer's metaphysical anti-realism is the consequence of his view that since mathematics is the outcome of the creative activity of the subject (see §4.8), it has nothing to do either with describing reality or with its applicability to the empirical sciences. Moreover, his anti-realism about truth is based on the idea that the only way of understanding mathematical truth is in terms of constructive provability. (Possibility p_{22}.)

Metaphysical anti-realism and anti-realism about truth are clearly compatible. In fact, in a situation in which there are neither natural numbers, nor a natural numbers structure, etc. it makes clearly sense to say that, given a number theoretical statement P, asserting 'P is true' can only mean asserting that P is provable, where 'P is provable' means 'I can effect a construction in my mind such that ...' (see §4.6 and following).

From the examination of the four above mentioned positions, I can conclude that the issues about metaphysical realism and realism about truth are independent of one another.

10 A new form of debate?

Having examined two different methodological approaches to the realism/anti-realism issue in the philosophy of mathematics, it is legitimate to ask how I should go about dealing with it myself. Should I agree with Carnap, or with Dummett, or with neither? Perhaps, some light can be cast on this question, if I discuss what has been learned so far from the study of the Carnapian and Dummettian indications on how the realism/anti-realism dispute should take place.

In analyzing Carnap's position with regard to the distinction he draws between *internal* and *external* questions of existence, it became immediately clear that, as Carnap asserts, it does not make sense to ask questions such as 'Are there numbers?' independently of a framework (mathematical theory). But, I argued, such a view, far from showing that 'questions concerning the existence or reality *of the system of entities as a whole*' are meaningless, points rather at the need to provide a context within which not only questions of existence, but any question at all, must be located to have meaning.

Indeed, we saw that external questions of existence periodically emerge in the history of mathematics, and are resolved by the mathematical community not by proving the correctness of a framework from higher principles, but through the testing of the framework in relation to its logical consequences, that is, in relation to whether these are acceptable or not. I considered, in particular, the problem about the existence of infinitesimals, and argued that the discussion which took place within the mathematical community about this issue revolved around a number of anomalies which were a consequence of their use in analysis. I also pointed out that such a debate, which had as protagonists two factions of philosophers, and mathematicians — the finitists and the infinitesimalists — factions which upheld different frameworks from one another, rather than being a mere rhetorical exercise, produced such compelling arguments, which eventually led the mathematical community to rid analysis of the infinitesimals.

But what conclusions should be drawn from these considerations? Well, the most immediate conclusion has to be that, with the overcoming of the problem represented by the incommensurability of frameworks, traditional and non-traditional external questions of existence must regain the status of legitimate philosophical questions.

Secondly, the study of what happens when external questions of existence are asked, or when the mathematical community adjudicates between competing mathematical frameworks, is something that must be firmly grounded in the history of mathematics. But, of course, this does not imply, as someone seems to think, that philosophy is irrelevant in the discussion of issues as the ones mentioned above.

Indeed, in the case of the question about the existence of infinitesimals, the history of mathematics reveals that the debate taking place between mathematicians was neither devoid of considerations concerning the so-called 'metaphysics of the calculus', nor was it immune to opinions about what mathematics *should* be like.

Thirdly, paying attention to such questions as 'What are the consequences of our thesis or our theory?', 'Are they all acceptable to us', 'How do mathematical theories change?', and 'Which criteria are used by the mathematical community to choose between different and opposing frameworks?' is crucial not only for giving a compelling account of how external questions of existence are posed, and resolved in mathematics, but also for addressing the problem of explaining the rationality of the change of mathematical theories, and that of comparing the way theory-change takes place in mathematics with the way in which theory-change takes place in the empirical sciences.

However, there is, perhaps, a deeper lesson to be learned from the several instances in which different and opposing frameworks are compared with one another by the mathematical community, and a choice is made between them based on rational criteria. Such a lesson consists in realizing that the abstract speculations concerning what mathematical theories are (or should be), speculations which bear no relevance to, or are irreconcilable with, mathematical practice are, to paraphrase Kant, empty.

On the other hand, it is also the case that considerations concerning the nature of mathematics which consist simply of an unreconstructed mass of data are bound to be blind.[45]

If we now shift the focus of attention from Carnap to Dummett, it must be said that one of the important things that emerge from the Dummettian way of setting up the realism/anti-realism dispute is the distinction between metaphysical realism and realism about truth.

Dummett has clearly shown that it is certainly possible to have a debate about whether a realist or an anti-realist theory of truth should be adopted for the statements belonging to the disputed class of a given mathematical theory, even though the adoption of a realist or an anti-realist conception of mathematical truth does not commit one to taking a particular position in metaphysics (and *vice versa*). A consequence of this last fact is that there does not seem to be a path from logic to metaphysics, and that therefore we must deal with the metaphysical issue of realism independently of a discussion of realism about truth.

The above considerations demand a radical change within the philosophy of mathematics concerning the best way of setting up the realism/anti-realism debate.

11 An appeal to history

From the abstract and unhistorical approaches of Carnap and Dummett, we need to turn to the idea that the most important presupposition for a sound and productive realism/anti-realism dispute about mathematics is neither a formal distinction between internal and external questions of existence nor a correct theory of meaning for the language of mathematics, but the understanding of the process according to which mathematical knowledge grows.

In fact, since mathematical knowledge grows either within a mathematical theory T — as a consequence of proving theorems of T — or, when a certain kind of revolutionary change takes place, as the result of the process according to which the mathematical community accepts some mathematical theories and rejects others, it follows that determining, for instance, which are the criteria adopted by the mathematical community to accept or reject a theory is of paramount importance for the realism/anti-realism debate in the philosophy of mathematics.

---

[45] A distinction drawn along similar lines may be found in [Lakatos, 1971], p. 102.

Indeed, if we were to discover that these are criteria which revolve exclusively around psycho-sociological or aesthetical considerations, this would be an important point in favour of metaphysical anti-realism about mathematics, because, for a mathematical theory T to be acceptable or to be better than another mathematical theory Q, the relation of T (and Q) with reality would not be an issue.

But if, on the other hand, such criteria appealed irreducibly to independently existing entities, whichever their ontological status might be, then this would be a very important point scored in favour of metaphysical realism about mathematics, because for a mathematical theory T to be acceptable, or to be better than another mathematical theory Q, the relation of T (and Q) with reality would be fundamental.

It is very important to notice that the study of the criteria according to which the mathematical community accepts or rejects mathematical theories, and the investigation of other phenomena accompanying this process of theory-choice such as: the presence and function of metaphysical assumptions of a realist or of an anti-realist character in mathematical theories; the shift in meaning of mathematical notions; the phenomenon of axiom introduction and axiom elimination; etc. has a decisive advantage with respect to Carnap's and Dummett's approaches to the realism/anti-realism debate. Such an advantage consists in the possibility of justifying a realist or an anti-realist position in the philosophy of mathematics through an appeal to objective criteria and phenomena which, besides being explanatory of the way the theory evolves and of how the knowledge produced by it grows, are deeply rooted in the history of mathematics.

However, at this point some of the readers whose philosophical sympathies side all with analytical philosophy might start to feel uneasy as they bring back to memory the famous passage from the *Grundlagen* in which Frege asks[46]

> Do the concepts, as we approach their supposed sources, reveal themselves in peculiar purity?

and replies[47]

> Not at all; we see everything as through a fog, blurred and undifferentiated. It is as though everyone who wished to know about America were to try to put himself back in the position of Columbus, at the time when he caught the first dubious glimpse of his supposed India. Of course, a comparison like this proves nothing; but it should, I hope, make my point clear. It may well be that in many cases the history of earlier discoveries is a useful study, as a preparation for further researches; but it should not set up to usurp their place.

---

[46][Frege, 1884], Introduction, p. vii.
[47][Frege, 1884], Introduction, pp. vii-viii.

11. AN APPEAL TO HISTORY

Indeed, such a disparaging view of the rôle that the history of mathematics has within the philosophy of mathematics has been a constant feature of much of analytical philosophy since the time of Frege. Such a view depends, among other things, on (a) Frege's belief in a mythology of concepts in their pure form the knowledge of which is achieved by 'stripping off the irrelevant accretions which veil [them] from the eyes of the mind'; and on (b) a too narrow opinion concerning the scope of the history of mathematics, which is seen by Frege simply as 'the history of earlier discoveries'.

Concerning point (a), we shall see in the discussion of Platonism contained in Chapter 3 that what I have called the 'mythology of concepts in their pure form' makes it impossible to work out a plausible mathematical epistemology.

As for point (b), we have already seen that, when we raise the gaze from the individual mathematical discoveries to the general dynamics regulating the growth of mathematical knowledge, the history of mathematics has a very important rôle to play in producing vital information concerning philosophical problems such as the realism/anti-realism debate in the philosophy of mathematics.

The use of the history of mathematics is, therefore, neither a way of '[betaking ourselves] to the nursery, [nor of burying ourselves] in the remotest conceivable periods of human evolution, there to discover, like John Stuart Mill, some gingerbread or pebble arithmetic!' It is, rather, a way which makes us discover and describe new phenomena which are relevant to some fundamental philosophical questions.

However, claiming that the history of mathematics must be at the heart of an acceptable way of debating the realism/anti-realism issue about mathematics, is not to say that the history of mathematics provides a philosophically neutral tribunal before which philosophical questions must be brought.

Indeed, I must distinguish very sharply my position from that of those who assert that since '... mathematics has a life of its own, independent of any philosophical considerations'[48] philosophy is irrelevant to mathematics and its history. I believe this view to be erroneous with regard to mathematical practice, as the case-history examined about the theory of infinitesimals shows, and especially misleading with regard to the history of mathematics.

For, the history of mathematics does not consist of a mere chronology of events, but of a rational reconstruction of the way mathematical knowledge grows; and this rational reconstruction is attained by means of selective attention paid to mathematical events which are distinguished into causes, effects, etc. according to a certain theory. The history of mathematics is, therefore, the outcome of the interpretation, and of the evaluation of certain

---
[48][Shapiro, 1997], Introduction, p. 7.

events obtained within the historiography of mathematics through the use of concepts and ideas which originate within the philosophy of mathematics. Therefore, I agree with Lakatos for whom:[49]

> ...the historiography of [mathematics] should learn from the philosophy of [mathematics] and *vice versa*

Although the assertion above has the ring of truth about it, and it is straightforward to see why the philosophy of mathematics should learn from the historiography of mathematics, it is, nevertheless, legitimate to ask what exactly does the philosophy of mathematics contribute to the historiography of mathematics.

A very compact answer to the problem of what does the philosophy of mathematics contribute to the historiography of mathematics is that the[50]

> ...philosophy of [mathematics] provides normative methodologies in terms of which the historian reconstructs 'internal history' and thereby provides a rational explanation of the growth of objective knowledge

To make it possible for the reader to understand the above answer I must explain what the internal and external histories of a mathematical theory are. The internal history of a mathematical theory is the account of how the mathematical theory has evolved in relation to the set of its problems.

A very important feature of the internal history of a mathematical theory is that it is normative, because the particular shape that this takes depends on the methodology you adopt to account for historical events.

To see this consider what happened in analysis when the infinitesimals were eliminated. If your methodology allows (normative aspect) theory-refutations, then you might explain the elimination of infinitesimals from analysis as an instance of the refutation of a mathematical theory. If, on the other hand, your methodology does not allow theory-refutations then you might say that the elimination of the theory of infinitesimals from analysis is an example of the elimination of misguided philosophical ideas which had been introduced within analysis when mathematicians had not yet hit against the genuinely mathematical concepts which give a correct account of differentiation, integration, continuity, etc.

It is this normative function performed by methodology in shaping up the internal history of a mathematical theory what prompts me to call such a methodology 'logic of mathematical discovery'. However, before proceeding any further, I must point out that my use of the term 'discovery' in relation to the normative methodology used in the historiography of

---

[49][Lakatos, 1971], p. 102.
[50][Lakatos, 1971], p. 102.

11. AN APPEAL TO HISTORY

mathematics is purely metaphorical, and should not suggest any realist bias in the understanding of the very concept of normative methodology.

What I mean by the 'external history' of a mathematical theory is the account of the socio-psychological conditions relevant to the occurrence of certain mathematical events. An example of external history was given in §1.6 where I described the way Leibniz's and Newton's ideas were received by the French Academy, and by philosophers like Berkeley.

With regard to the rôle of external history, I must point out that[51]

> ...any rational reconstruction of history needs to be supplemented by an empirical (socio-psychological) 'external history'.

In fact, only the external history is capable of explaining why a mathematical theory was developed in a particular country at a particular time rather than in other countries; the speed with which certain ideas were accepted; etc. by making explicit the socio-cultural factors which, together with the demands made by the economy on mathematics in terms of technological applications, play the very important rôle of catalysts.

But, at this point, having clarified the concepts of internal and external history, it is important, before bringing this section to a close, to address the following question. If the philosophy of mathematics contributes normative methodologies to the historiography of mathematics, normative methodologies which are used by the historian to reconstruct the internal history of a mathematical theory, what happens when two different normative methodologies come into conflict with one another?

The answer to the question above is that[52]

> ...two competing methodologies can be evaluated with the help of (normatively interpreted) history

in the same way in which two different frameworks can be compared with one another. In other words, as in the case concerning the normative methodologies (inductivism, falsificationism, etc.) adopted within the empirical sciences to write their internal histories, there has been a rational debate, a rational debate which has led to the rejection of some of them, in the same way, in the case offered by mathematical theories, it is possible to come to a rational decision about which among two competing normative methodologies is the best by examining the consequences deriving from their adoption.

---

[51][Lakatos, 1971], p. 102.
[52][Lakatos, 1971], p. 102.

12 Naturalism about mathematics?

Although my view of the method that must be adopted to decide the realism/anti-realism dispute in the philosophy of mathematics should be sufficiently clear by now, it is important to distinguish my position on this issue from what has come to be known as naturalism about mathematics.

According to Penelope Maddy, for a naturalist about mathematics:[53]

> ...mathematical methodology is properly assessed and evaluated, defended or criticized, on mathematical, not philosophical (or any other extra-mathematical) grounds.

Moreover, she adds,[54]

> ...naturalism about mathematics differs markedly from naturalism about science in its treatment of philosophical considerations. Consider, for example, the typically metaphysical claim that mathematical objects exist objectively and non-spatiotemporally. The mathematical naturalist holds these issues to be external to mathematics proper and thus irrelevant to methodological decision-making, but the analogous claim about physical objects — that they exist objectively and spatiotemporally — is part and parcel of scientific thinking. This is not true of the corresponding mathematical questions, which is why the mathematical naturalist undertakes to eliminate them from methodological arguments.
>
> For these reasons, the mathematical naturalist pursuing questions of methodology ignores traditional philosophical questions such as 'are mathematical things objective or subjective?', 'is their existence dependent on our theories or definitions?', 'are mathematical objects incomplete?', 'are they more like fictional objects or physical objects?', 'are the axioms true in the real world of sets?' and so on.

Now, even though my suggestion of investigating how mathematical knowledge grows, with the view of settling the realism/anti-realism debate about mathematics, is, indeed, a suggestion to engage in an historical study of mathematical methodology, this is a genuinely philosophical, and not mathematical, enterprise whose success can be critically assessed on the basis of the accuracy of its rational reconstructions.

An example of what I mean by this is given in Chapter 7 of this book. There I apply an unashamedly philosophical normative methodology — a modified version of Lakatos's Methodology of Scientific Research Programmes (MSRP) — to the historiography of Cantor-Zermelo set theory. This is done to produce a reconstruction of the internal history of Cantor-Zermelo set theory from which it should clearly emerge how mathematical knowledge grows within Cantor-Zermelo set theory.

Concerning the peculiarities of my approach, observe that, first, the application of some kind of MSRP to Cantor-Zermelo set theory is a genuinely philosophical operation, because it is neither aimed at solving a mathematical problem nor at developing a mathematical theory, but is meant to

---

[53][Maddy, 1998], §1, p. 164.
[54][Maddy, 1998], §3, pp. 172–173.

respond to a typically epistemological question related to the way mathematical knowledge grows.

Secondly, the correctness of the rational reconstruction of the internal history of Cantor-Zermelo set theory effected through the application of some kind of MSRP to Cantor-Zermelo set theory, is something that can be rationally tested against both the evidence offered by how Cantor-Zermelo set theory has evolved, and other rival reconstructions offered.

Thirdly, in contrast with the mathematical naturalist's rejection of questions like 'Are mathematical things objective or subjective?', 'Is their existence dependent on our theories and definitions?', etc. because irrelevant to his investigations of mathematical methodology, we shall see that the rational reconstruction of the internal history of Cantor-Zermelo set theory will reveal, among other things, that Cantor-Zermelo set theory is based on a number of very strong metaphysical assumptions which have a realist connotation.

Lastly, the aim of studying mathematical methodology through the production of rational reconstructions of the internal history of mathematical theories, rational reconstructions obtained using some normative methodology as a schema for the interpretation of historiographical data, is not that of providing an accurate description of mathematical thinking, but that of studying how mathematical knowledge grows, that is, if mathematical knowledge always grows cumulatively or whether there are also times when it grows in a revolutionary way, e.g., through the rejection of old theories which are replaced by new ones, the change of meaning of important mathematical concepts such as that of axiom, proof, etc.

The considerations above show in a very clear way that my approach to the study of mathematical methodology is very different from that of the naturalist about mathematics, and that, for me, although 'the goal of mathematical epistemology is not that of producing a validation of non-empirical science', but the more modest one of producing correct descriptions of mathematical knowledge, mathematical epistemology cannot be naturalized in the sense that, even if mathematical epistemology has to be grounded in considerations regarding the history of mathematical theories, is in no way reducible to them.

CHAPTER 2

ARGUMENTS FOR REALISM IN MATHEMATICS

1 Introduction

The traditional discussions over realism in mathematics have an exclusive metaphysical flavour, and metaphysical realism about mathematics goes back a long way in the history of philosophy as is shown by the fact that Pythagoras and Plato are two of the earliest and foremost representatives of this position.

The method I have chosen to provide a review of the case made by some philosophers in favour of realism in mathematics consists in analyzing the traditional realist replies to the following questions: (i) which arguments are there in support of realism in mathematics? (ii) what is the nature of mathematical reality? and (iii) how do we acquire knowledge of mathematical reality?

In the present chapter I am neither going to concentrate on the various views concerning the nature of mathematical reality (question (ii)); nor on the several accounts offered by realists of the acquisition of mathematical knowledge. These topics will be dealt with in Chapter 3. I shall, instead, content myself with accomplishing two preliminary tasks: (a) showing that the question about mathematical realism can be reduced to that concerning whether or not one should be realist with regard to set theory; and (b) examining some of the main arguments in favour of set-theoretical realism (question (i)).

The importance, and preliminary character, of the first task reside in its simplifying function. In fact, if it were possible to accomplish it, instead of having to make a case in favour of metaphysical realism for every single mathematical theory, it would become possible to argue in favour of metaphysical realism in mathematics once and for all simply with regard to set theory.

The importance of the second task consists in producing arguments in favour of metaphysical realism which, being independent of any particular mathematical ontology, may, on the one hand, prevent metaphysical realism's suffering from the faults connected with the implausibility of any particular suggested ontology; and, on the other hand, provide a solid foundation for a subsequent investigation of the nature of mathematical reality.

2 From realism in mathematics to set-theoretical realism

Before set theory, philosophers of mathematics assigned to various mathematical theories different rôles at different times. Pre-Gaussian mathematics, for example, held Euclidean geometry as the model of what a mathematical theory should be like, and looked at number theory and analysis as the poor relations.

This situation had been established for a long time in the Euclidean tradition where the very concept of number receives a geometrical interpretation. A number is, in fact, represented by Euclid as the length of a line segment, and the product of two numbers as the area of a rectangle.[1]

In the post-Gaussian era the tables were turned. Euclidean geometry became one of several possible types of geometries that could be studied, and number theory was, indeed, crowned as the queen of mathematics.

As is well known, this dramatic change was caused by several factors such as the introduction of systems of non-Euclidean geometry and the arithmetization of analysis. This change then culminated in the axiomatization of number theory, and by number theory receiving from Frege and Russell the exhalted status of being part of logic.

Given the change of priority between mathematical theories, a situation which has occurred several times in the history of mathematics, arguing for realism in mathematics should then mean arguing for realism with regard to the entities described by every single mathematical theory. A very difficult task to accomplish!

However, with the advent of set theory the situation changed once more. Such a change was the consequence of the discovery that the known mathematical theories can be expressed within set theory.

---

[1] Some of the definitions from Book VII of the *Elements* are particularly important with regard to this question. I shall mention here only two of them ([Euclid, *The Elements*], **vol. 2**, Book VII, p. 278):

Definition 2.1 (16). And, when two numbers having multiplied one another make some number, the number so produced is called **plane**, and its **sides** are the numbers which have multiplied one another.

Definition 2.2 (17). And, when three numbers having multiplied one another make some number, the number so produced is called **solid**, and its **sides** are the numbers which have multiplied one another.

3. THE OBJECTIVITY ARGUMENT

For some, this result is philosophically very important, because it highlights the fact that set theory is more fundamental than any mathematical theory expressible within it, in the sense that all the theories which are expressible in set theory are 'really' about sets.

For others, this only shows that the language $\mathcal{L}_T$ of a mathematical theory T is translatable into the language of set theory $\mathcal{L}_\epsilon$ in such a way that the structure of T is preserved under translation, and the translation of a theorem of T is a theorem of set theory.

Moreover, they would add, this translation of, say, number theory into set theory hides number theoretical facts, because doing number theory in a full set-theoretical framework increases the complexity of computations with respect to those taking place in traditional arithmetic. This is the case, because the set-theoretical formulae which represent numbers are more complex than the corresponding numerals, and the set-theoretical algorithms are much more complex than the number-theoretic ones they are meant to simulate; and, of course, any increase in complexity not only decreases the efficiency of the algorithms used, but also undermines surveyability.

To give an example of what they mean, they would then challenge the set-theoretic enthusiast to carry out, using von Neumann ordinals, the simplest arithmetical operations such as $350 + 7754$, and compare the complexity of the set-theoretical computation with that of the ordinary arithmetical one.

Whichever of these two factions is right about the interpretation of the philosophical significance of these 'expressibility results', the fact remains that arguing in favour of set-theoretical realism would be arguing in favour of mathematical realism in general, because since for any theorem belonging to a mathematical theory T expressible in set theory there is a corresponding set-theoretical theorem, we could claim that any theorem of T would have to be true not simply because it has been proved in T, but also because it describes properties of sets.

3 The objectivity argument

According to the believers in the correctness of the objectivity argument: (1) it is a fact that, in contrast with statements such as (a) 'Good is an end in itself' or (b) 'Michelangelo is a better painter than Picasso', statements like (c) '$2+3=5$' or (d) 'If A is a set with n elements, the power set of A has 2^n elements' command universal agreement among those who know arithmetic and set theory; and (2) the best explanation we have for the objectivity of statements such as (c) and (d) is the existence of set-theoretical entities corresponding to the singular terms occurring in (c) and (d).

To this the set-theoretical anti-realist can reply that, to explain the objectivity of statements like (c) and (d), we do not need to conjure up or postulate the existence of objects described by such statements. For, the objectivity of (c) and (d) more simply derives from the proofs of these statements.

Notice that the set-theoretical anti-realist's explanation of the objectivity of mathematical statements appears to be particularly strong in the mouth of the constructivist, for whom mathematical proofs are only those which are conceivable as procedures of verification. Indeed, what is more objective than a procedure of verification?

In spite of its immediate appeal, this is not a knock-out objection, since the set-theoretical realist can reply that, first, basing mathematics on constructive proofs does not of itself guarantee mathematical objectivity, above all in a situation in which there is no common agreement among the practitioners on what we should consider to be a constructive proof.[2]

Secondly, if we analyze a given deduction going from the proved statement C back to the axioms involved in the proof, we will realize that it is often the case that some such axioms are not self-evidently true, and that, consequently, our proof, rather than justifying the assertion of C, can only justify the assertion of the conditional statement $(A_1 \wedge \cdots \wedge A_n) \to C$, where $A_1, \ldots, A_n$ are the axioms used in the proof of C.

Moreover, since, as a consequence of Gödel's Incompleteness Theorems, we cannot prove the consistency of the set of axioms of set theory within theories which are provably beyond any suspicion of inconsistency, it follows that we cannot produce a coherentist justification for the assertion of C either, that is, we cannot produce a justification for the assertion that C can be derived from a consistent set of axioms.

However, the set-theoretical realist would continue, when we study the history of set theory, we observe that the adoption of some of its axioms is not the outcome of arbitrary choice, but the result of the application to the axiom candidates of a number of rational criteria which are not reducible to considerations which exclusively involve self-evidence, consistency, simplicity, or aesthetical value. And if we examine some such criteria — like success in solving open problems, fruitfulness in unforeseen applications to various mathematical theories and to the empirical sciences, etc. — we realize that these criteria must ultimately rest on the existence of a set-theoretical reality which, therefore, imposes upon us the axiom candidates which satisfy them.

In support of the correctness of the thesis above, the set-theoretical realist would then treat us to a rational reconstruction of particularly significant case-histories exemplifying the process according to which axioms are chosen in mathematics.

---

[2] See on this §1.8 and §§4.8–4.12.

4 Two case-histories

In what follows in this section I am going to consider two such case-histories. The first is about the introduction of the Axiom of Choice (**AC**), whereas the second concerns the elimination of the Axiom of Reducibility (**AR**).

It is particularly important to consider both a case of axiom introduction and a case of axiom elimination to show that mathematical reality is the decisive factor behind the two different aspects of the process of choosing the axioms of a mathematical theory.

The introduction of the Axiom of Choice

In an article published in 1883 Cantor introduced the Well-Ordering Principle which states that: any set A — actually Cantor speaks of *well-defined sets*[3] — can be well-ordered.[4]

The Well-Ordering Principle plays a very important rôle in Cantorian set theory, because it provides the presupposition for the comparability of sets in relation to their size.[5] It is, therefore, fundamental to establish the meaningfulness of the very notion of *size* of a set as what can be determined not by the operation of counting but by relative comparison.[6]

At the time of the introduction of the Well-Ordering Principle Cantor believed that this was a law of thought[7] and that, as such, was not in need of a proof. However, his attitude towards the status of the Well-Ordering Principle subsequently changed. Cantor came to believe that the Well-Ordering Principle could not have the status of self-evident truth, typical of a law of thought, and that, therefore, the acceptance of such a principle was going to be subject to a proof.

This change was perhaps, as Moore says in his book, a consequence of

---

[3] Cantor clarifies his notion of well-defined set in the following passage which occurs in [Cantor, 1882]. The translation is given in [Hallett, 1984], p. 45:

> I call a manifold (a totality, a set) of elements which belong to some conceptual sphere well-defined, if on the basis of its definition and as a consequence of the logical principle of excluded middle it must be seen as *internally determined* both whether some object belonging to the same conceptual sphere belongs to the imagined manifold as an object or not, *as well as* whether two objects belonging to the set are equal to one another or not, despite formal differences in the way they are given.

[4] A partially ordered set A is *well-ordered* if any non-empty subset B of A contains a least element. Cantor first mentions the Well Ordering Principle in [Cantor, 1883].

[5] The Well-Ordering Principle is equivalent to the Trichotomy law for cardinal numbers: if m and n are cardinal numbers then either $m < n$ or $m = n$ or $m > n$.

[6] On Cantor's views concerning the Well-Ordering Principle and further developments affecting the relevance of such a Principle see: [Hallett, 1984], Ch. 3, §3.5 and [Moore, 1982], §§1.5–1.6.

[7] See [Moore, 1982], Ch. 1, p. 42.

44 ARGUMENTS FOR REALISM IN MATHEMATICS

the scepticism that several mathematicians had towards its truth.[8] It is, in fact, highly counter-intuitive, in the presence of sets like $\mathbb{R}$, which are not well-ordered under their natural ordering, and in the absence of a general procedure of choice, to think that, given a set with a very large cardinality, this can be ordered in such a way that every one of its non-empty subsets has a least element.

The Well-Ordering Principle became eventually a theorem when Zermelo provided a proof for it in 1904,[9] a proof which depended on the a principle later called 'Axiom of Choice' (**AC**).[10]

The introduction of the Axiom of Choice, made by Zermelo to prove the Well-Ordering Theorem, was not dependent on this principle being recognised as self-evidently true.

In fact, as Moore puts it:[11]

> ...a majority of those mathematicians who expressed an opinion about Zermelo's proof [of the Well-Ordering Theorem] rejected it unequivocally, and usually dismissed the Axiom of Choice in particular.

Such a negative reaction of a large part of the mathematical community to the introduction of the Axiom of Choice was essentially based on two problems connected with it. The first was represented by the (apparent) mathematical meaninglessness of the idea, which came with **AC**, of an infinite number of arbitrary choices. And the second was, instead, related to a number of perplexing consequences of the introduction of **AC** such as the Tarski-Banach paradox,[12] and the existence — proved by Vitali in 1905 — of non-Lebesgue measurable sets of reals.

The first problem was solved, for most mathematicians, when the purely existential character of the Axiom of Choice became clear:[13]

> In fact, the axiom does not assert the possibility (with scientific resources available at present or in any future) of *constructing* a selection-set; that is to say, of providing a rule by which in each member s of t a certain member of s can be named. On

---

[8] See [Moore, 1982], pp. 50–51.
[9] See [Zermelo, 1904].
[10] Zermelo's original formulation of this axiom was: To each subset M' one associates any element m'_1 which occurs in M' itself and which may be named the 'distinguished' element of M'. See [Moore, 1982], Ch. 2, §2.2, p. 90.
[11] See [Moore, 1982], pp. 142–143.
[12] The Tarski-Banach paradox is given by the following theorem:

Theorem 4.1. A closed ball U can be decomposed into two disjoint sets, $U = X \cup Y$ such that $U \approx X$ and $U \approx Y$.

in [Jech, 1973], Ch. 1, §1.3, p. 3. In other words, there is a way, using the Axiom of Choice, of decomposing a sphere U into two spheres X and Y each of which is congruent to U. See also [Moore, 1982], Ch. 3, §3.7 and Ch. 4, §4.11; and [Wagon, 1985].

[13] In [Fraenkel et alii, 1973], Chapter II, §4.4, p. 68.

4. TWO CASE-HISTORIES

the contrary, providing such a rule would mean obtaining the respective subset of $\bigcup t$ by the axiom of subset, without involving the axiom of choice. The latter just maintains the *existence* of a selection set, i.e. the non-emptiness of the outer product[14] πt (whose existence is guaranteed without our axiom).

Concerning the second problem, it must be said that, first, an agreement was reached among mathematicians on the idea that Vitali's use of **AC** showed the limitations of Lebesgue measure — such an agreement was, indeed, one of the driving forces behind the development of measure theory. And, secondly, that the so-called Tarski-Banach paradox (see Theorem 4.1, footnote 12), and a whole host of similar results, do not represent a real challenge for **AC**, because these results, which we can prove using **AC**, are neither false statements like '0 = 1' nor contradictions. They are, rather, statements which are presently found repugnant by most mathematicians, perhaps, in the same way in which Saccheri found repugnant the theorems of hyperbolic geometry he proved.

'However', the set-theoretical realist would remark, 'it is extremely instructive to see that, in spite of the two above mentioned problems, and an initial climate of open hostility following its introduction, the Axiom of Choice has forced itself upon us as being true' — to paraphrase the Gödel of 'What is Cantor's continuum problem?'. But what caused such a dramatic change of attitude of the mathematical community towards the Axiom of Choice?

Well, certainly, the ability to provide a satisfactory answer to the first of the two above mentioned problems, and an **AC**-friendly reinterpretation of the phenomena illustrated in the second, played a part in this, but the most decisive factors behind the radical shift in the mathematicians' attitude towards **AC** were, in increasing order of importance: (1) the proof of the relative consistency of **AC** with respect to **ZF**;[15] (2) the proof of the independence of **AC** from the axioms of **ZF**; and (3) the vast, profound, and entirely unforeseen mathematical importance of **AC** and of statements equivalent to it.

In fact, first, the relative consistency of **AC** with respect to **ZF**, i.e., that Con (**ZF**) $\Rightarrow$ Con (**ZFC**), shows that there is nothing to fear from adding **AC** to **ZF**.

---

[14]

Definition 4.1. A disjointed set t is a set whose members are pairwise disjoint.

Definition 4.2. If t is a set, the *outer product* πt is the set whose members are just those sets which contain a single member from each member of t.

[15]By **ZF** I mean the set of all the Zermelo-Fraenkel axioms minus **AC**, whereas **ZFC** is **ZF** + **AC**.

Secondly, the independence of **AC** from **ZF** speaks of the fundamental nature of **AC** as a genuine axiom candidate able to produce a safe and genuine extension of **ZF**.

Thirdly, and most importantly, in the light of the two above mentioned points, the unforeseen, at the time of its introduction, mathematical relevance of **AC**, and of statements equivalent to it, in the most different branches of mathematics — set theory, analysis, topology, algebra, etc.[16] — provides clear evidence in favour of the idea that **AC**, far from being a purely *ad hoc* hypothesis devised to 'fix' an important result in set theory — the Well-ordering Theorem — is a new principle whose truth is neither self-evident nor provable (in **ZF**), but is forced upon us, in the face of strong opposition, by mathematical reality (what else?) through the important, and sometimes surprising, consequences of its introduction.

The considerations made in this third point may be summed up in a very compact way by the well known Biblical principle:[17]

By their fruit you will recognize them.

The elimination of the Axiom of Reducibility

Let us now analyze a negative case of axiom choice: axiom elimination. The case in question is represented by Russell's Axiom of Reducibility.

According to the Russell of *Principia*, at the root of *all* the paradoxes which beset naïve set theory there are violations of the so-called Vicious Circle Principle according to which[18]

Principle 4.1 (Vicious Circle Principle). Whatever involves *all* of a collection must not be one of the collection.

Such a principle boils down to a restriction imposed upon the language of set theory which, in particular, condemns as nonsensical expressions like $x \in x$ and, consequently, also their counterpart $x \notin x$. As is well known, these are expressions which are key components of the breeding ground of the paradoxes of set theory.

Following the dictates of his Vicious Circle Principle, Russell (and Whitehead) devised the Ramified Theory of Types (**RTT**), which is able to bar as nonsensical not only expressions crucially involved in paradoxes like Burali-Forti's, Cantor's and Russell's, but, through the ramification of types into orders, also expressions at the root of paradoxes like Richard's, Berry's and Grelling's.

---

[16] To those interested in a discussion of the mathematical relevance of the Axiom of Choice I recommend reading [Moore, 1982], Ch. 1, §1.7; Ch. 3, §3.5; Ch. 4; and [Rubin & Rubin, 1963].

[17] Matthew 7:16.

[18] [Whitehead & Russell, 1962], Ch. II, §I, p. 37.

4. TWO CASE-HISTORIES 47

However, as a consequence of the non-cumulative nature of Russell's types,[19] of the ramification of these into orders and of the intensional character of Russell's logic two major problems arose: (1) the impossibility of defining entities such as the least upper bound of a given set (and the greatest lower bound) and, therefore, the impossibility of using common proof-techniques within analysis (Dedekind's cuts for the definition of real numbers, proof of the Intermediate Value Theorem, etc.); and (2) the impossibility of giving an adequate definition of identity.

To see the genesis of the first problem in Russell's system, let us, for the sake of clarity of the exposition, adopt the extensional notation of contemporary set theory.

Given a set Z bounded from above and the set $Z^\uparrow$ of upper bounds of Z, the least upper bound of Z is usually defined as the element b of $Z^\uparrow$ such that $b \leq x$, for any $x \in Z^\uparrow$.

If we read the above definition with some care, we realise that the least upper bound b of Z is defined by appealing to the totality $Z^\uparrow$ of which b is an element or, to put it in another way, b and the elements of $Z^\uparrow$ ought to have two different logical orders, which is absurd, because it would follow that b ought to have a logical order different from itself. Therefore, we cannot define the least upper bound in **RTT**.

To illustrate the second problem from which Russell's Ramified Theory of Types suffers, we better switch to the intensional notation of *Principia*, which is one of the main causes of the problem, and let Russell and Whitehead formulate it themselves:[20]

> It is plain that, if x and y are identical, and ϕx is true, then ϕy is true. Here it cannot matter what sort of function $\phi \hat{x}$ may be: the statement must hold for *any* function.[21] But we cannot say, conversely: "If, with all values of ϕ, ϕx implies ϕy, then x and y are identical"; because "all values of ϕ" is inadmissible. If we wish to speak of "all values of ϕ", we must confine ourselves to functions of one order. We may confine ϕ to predicates, or to second-order functions, or to functions of any order we please. But we must necessarily leave out functions of all but one order. Thus we shall obtain, so to speak, a hierarchy of different degrees of identity.

Russell's logic is intensional and not extensional and, therefore, it does not admit of a quick and satisfactory definition of identity as is, for instance, provided in set theory.[22] Moreover, the impossibility, given an x, of talking

---

[19] We have *cumulative* types when, given any two ordinals α, β such that $\alpha \leq \beta$, $T_\alpha \subseteq T_\beta$ (T_α and T_β are types).

[20] See: [Whitehead & Russell, 1962], Ibid., §VI, p. 57.

[21] In the language of *Principia Mathematica* a distinction is drawn between the symbols '$\phi\hat{x}$' and 'ϕx'. $\phi\hat{x}$ stands for the propositional function '$\hat{x}$ is ϕ', whereas ϕx is *any* value of $\phi\hat{x}$.

[22] Given two sets A and B, we say that $A = B$ if and only if $A \subset B$ and $B \subset A$.

about all the ϕ's of x makes it really impossible to provide a satisfactory definition of identity.

Faced with these very serious difficulties Russell resorted to the introduction, as an axiom, of a very counter-intuitive principle, which seemed to do the job of reconciling the two problems analysed above with the Ramified Theory of Types. This principle was the Axiom of Reducibility which states that:[23]

> ...given any function $\phi\hat{x}$, there is a formally equivalent *predicative* function, i.e. there is a predicative function[24] which is true when ϕx is true and false when ϕx is false.

However, the Axiom of Reducibility has been rejected, despite the fact that it seems to be doing the job of solving two of the major difficulties generated by Russell's Ramified Theory of Types. These are some of the reasons for the rejection of **AR**.

First, the axiom is not simply counter-intuitive, but it seems a purely *ad hoc* postulation which, by a sort of miracle, produces the desired result.

Secondly, since **AR** lacks self-evidence, universality, necessity, etc. it follows that **AR** cannot be considered to be a principle of logic and, consequently, it should not be accepted by a logicist as an axiom of **RTT**, because it cannot be used as a means for the realization of a logicist programme such as that of *Principia*.

Thirdly, Russell's theory of types can be *simplified*, as shown by Ramsey, retaining the types-distinctions, but dispensing with the ramifications of types into orders (such a simplification renders the Axiom of Reducibility redundant[25]).

Fourthly, type-theories can be rendered even more simple and effective than Ramsey-style theories of types through Zermelo's introduction of cumulative types and the passage from a logic of intensions to a logic of extensions (set theory).[26]

---

[23] See: [Whitehead & Russell, 1962], Ibid., p. 56.

[24] A 1-place function ϕ is *predicative* if the order of ϕ is the next above that of its argument. If ϕ is an n-place function and the highest order of its arguments is k then ϕ is predicative if and only if the order of ϕ is $k+1$.

[25] Ramsey, following Peano, divides the set of paradoxes into two classes A and B. Paradoxes belonging to class A are those which would occur in a *naive* logical or mathematical system itself (Russell's paradox, Burali-Forti's, etc.), whereas paradoxes belonging to class B are those which cannot be stated in logical terms alone, but contain reference to thought, language, etc. (the *liar*, Grelling's paradox of 'heterological', Richard's paradox, etc.). He then proceeds to show that the type-structure is sufficient to eliminate paradoxes of class A and argues that, as far as the foundations of mathematics are concerned, we should not bother with those belonging to class B, which we might still find interesting for other reasons. See [Ramsey, 1925] in [Ramsey, 1978].

[26] In such systems we do not need the Russellian notion of order of a propositional

Fifthly, the Axiom of Reducibility is of no *mathematical consequence* in the sense that no result of analysis, algebra, etc. seems to be strictly dependent on it. Indeed, **AR** has a purely structural value confined to Russell's (and Whitehead's) system of the Ramified Theory of Types. It performs the same function as that of the hypothesis of the existence of epicycles in Ptolemaic astronomy: it is an *ad hoc* hypothesis introduced to eliminate aberrations.

If we, now, compare what I have said earlier in the analysis of the introduction of the Axiom of Choice with the considerations made concerning the Axiom of Reducibility some light might be shed on the process of axiom introduction.

Although the Axiom of Choice and the Axiom of Reducibility are both statements which postulate the existence of certain entities and which are not self-evidently true, the Axiom of Reducibility follows, contrary to the Axiom of Choice, the destiny of an obsolete theory — **RTT** — because it does not describe an aspect of mathematical reality.

5 The identity and inexhaustibility arguments

Another traditional argument in favour of set-theoretical realism is based on the idea that a necessary condition for believing in the existence of, say, the von Neumann ordinal $\{\emptyset, \{\emptyset\}\}$ is that we have identity conditions for $\{\emptyset, \{\emptyset\}\}$. In other words, if the symbol '$\{\emptyset, \{\emptyset\}\}$' refers to an object then, given any set A, it must be (in principle) possible to decide whether or not $A = \{\emptyset, \{\emptyset\}\}$. This is a thought common to various realist philosophers of mathematics such as Cantor[27] and Frege.[28] And, indeed, within systems like Zermelo-Fraenkel (**ZFC**) the existence of such identity conditions for sets is guaranteed by the so-called 'Axiom of Extension'.

---

function and, once again, the Axiom of Reducibility is not necessary. In these systems the notion of equality is not a problem at all but it can be defined in the usual set-theoretical way (see footnote n. 22). Moreover, Ramsey had already criticised the system of *Principia Mathematica*, because it denied the extensional nature of mathematics:

> The possibility of indefinable classes is an essential part of the extensional attitude of modern mathematics which we emphasised in Chapter I, and that it is neglected in *Principia Mathematica* is the first of the three great defects in that work. The mistake is made not by having a primitive proposition asserting that all classes are definable, but by giving a definition of class which applies only to definable classes, so that all mathematical propositions about some or all classes are misinterpreted.

The text occurs in [Ramsey, 1925] ([Ramsey, 1978], pp. 174–175).

[27] See footnote 4 in this chapter.

[28] [Frege, 1884], §62, p. 73:

> If we are to use the symbol a to signify an object, we must have a criterion for deciding in all cases whether b is the same as a, even if it is not always in our power to apply this criterion.

But surely, the fact that we have identity conditions for von Neumann ordinals, for example, is not sufficient to show that they exist. We could, in fact, think — like the formalists — that '**ZFC**' is just the name of a game played with certain finite strings of symbols used in such-and-such a way, etc. — one of these finite strings of symbols being {∅, {∅}} — and that these are finite strings of symbols for which we have identity conditions, but which do not refer to independently existing objects.

From this we can conclude that the force of the argument from identity is only *negative*, in the sense that the argument from identity does not provide a justification for the belief in certain entities, but only shows to be unwarranted the belief in entities for which it is not possible to give identity conditions. This is expressed in a very clear way by the well known Quinean quip 'No entity without identity'.

At this point our set-theoretical realist must think again about what might be a sufficient condition for believing in the existence of sets. And, in accepting the challenge of the anti-realist, the set-theoretical realist can, now, say that a sufficient condition for believing in the existence of sets can be produced, if we succeed in showing that the objects of investigation of set theory share with concrete objects certain characteristics typical of entities which exist independently of whether we are thinking about them or not. One most likely candidate for this rôle is the characteristic of concrete objects known as inexhaustibility.

Indeed, when you consider an animal or a rock, the amount of information you obtain about these entities is ever increasing, and often changing also with regard to the quality of information you first had about them. For, when we deal with entities which exist independently of whether we are thinking about them or not, there is always the possibility that we could be wrong in describing and/or classifying them.

A typical example of this phenomenon is represented by how what we know about whales has changed over time. As is well known, until not very long ago, whales were considered to be large fish. But nowadays, as a consequence of a great increase in the knowledge of their anatomy and physiology, these animals are no longer classified as fish but as mammals.

The situation concerning inexhaustibility undergoes a radical change when we ask whether the pieces of a game or the characters of a fiction story have such a property. To see this consider chess. In the game of chess, once you have defined how the King moves and what constitutes winning a game, there is nothing more to know about the King. And, even though there is much to study about chess end-games, and about how the King should be used in certain other circumstances, your actual knowledge about the chess piece itself is all contained in the rules of chess.

5. THE IDENTITY AND INEXHAUSTIBILITY ARGUMENTS

The reason for this is that, since the meaning and the conditions of correct use of the rules of chess are entirely established by convention, the rules of chess cannot be wrong because they produce an incorrect description and/or classification of the King, Queen, Bishop, etc.

Similar considerations apply to the characters of a fiction story. In fact, if we consider the case of fictional characters like Ulysses or Æneas, all there is to know about them can be 'exhausted' through the analysis of the information concerning these characters contained in the works of Homer and Virgil, because being, as I said, Ulysses and Æneas fictional characters, it makes no sense to doubt the correctness of the descriptions offered of Ulysses and Æneas as, respectively, 'The son of Laertes' and 'The son of Anchises and Aphrodite'.

If we, now, examine the objects of study not just of set theory, but of mathematics in general, with regard to inexhaustibility, we quickly realize that the situation we have in mathematics differs radically from that of a game or of a fiction story. In fact, if we, for instance, consider the concept of number, a study of the history of mathematics reveals that this concept has undergone many changes from the time of its first documented introduction within Western mathematics, and that the amount of information available to us about numbers has enormously increased since then.

To have just a glimpse of this, consider that the early Pythagoreans and Euclid had a 'geometrical' concept of number. Indeed, the Pythagoreans[29]

...did not really distinguish numbers from geometrical dots

and, for Euclid, numbers were thought to be the lengths of line segments (see §2.2).

However, the subsequent extensions of the number field, and the intense investigation of its algebraic, order-theoretical, and topological properties, besides causing dramatic theoretical changes in the study of numbers, have profoundly affected the very concept of number.

A typical example of this last phenomenon is that, as a consequence of the investigations of Cantor, Dedekind, Frege, Gauss, and Peano, the concept of number has become independent of geometrical intuition, eventually assuming a set-theoretical connotation according to which numbers must be conceived as certain kinds of sets.

Of course, this situation is not confined to number theory, but, being common to all mathematical theories, appears to provide support in favour of the idea that there is such a thing as mathematical reality.

---

[29] See [Kline, 1972], Ch. 3, §5, p. 29.

6 The theory acceptability argument

If we study the history of mathematics, we realize that there are times when new mathematical theories appear on the scene. And, it is interesting to notice that, when this happens, some of them are accepted by the mathematical community, and others are not.

As I argued in §1.11, the study of the criteria according to which the mathematical community accepts/rejects a mathematical theory T should prove relevant to the realism/anti-realism debate about mathematics.

Indeed, consider that when a new mathematical theory T is given, for T to be accepted by the mathematical community, T must, among other things, satisfy the 'criterion of the model', that is, it must be shown that T has a model in an already accepted mathematical theory T'.

If we ask an anti-realist to explain the reason for this, he will, perhaps, say that such a requirement is a response to the need to provide an assurance concerning the consistency of T, even though this is only relative to an already accepted theory T'.

But, this reply is somewhat unsatisfactory, if we consider that anomalies and inconsistencies have never restrained the mathematical community from accepting, and developing theories. Classical examples of this phenomenon are provided by the history of mathematical analysis, and by that of set theory.

On the other hand, the realist's explanation, according to which, for a theory T to be accepted by the mathematical community, one of the things that must be shown is that the domain of the intended interpretation of the language of T is non-empty, and that such a domain has as elements entities which have already been accepted as mathematical entities (or sets of entities which have already been accepted as mathematical entities, etc.), is much more plausible than the anti-realist's.

For the realist's explanation:

(1) provides a good reason why the statements of the new theory have mathematical content: these statements, and in particular those of existence, can be shown to be true of (already accepted) mathematical entities;

(2) gives a very plausible answer to the epistemological question concerning how we could imagine/represent what our theory is about;[30]

---

[30]This problem was particularly acute when the non-Euclidean geometries were introduced, so much so that, for Kant and Frege (!)

...it is impossible to form a mental representation of space-relations at variance with Euclid's geometry. ([Helmholtz, 1878], p. 686.)

As is well known, the problem of how we could imagine/represent the objects studied by hyperbolic and elliptic geometry was positively solved by the discovery of Euclidean models of these two systems of non-Euclidean geometry. (See next section for more

7. SOME CASE-HISTORIES   53

(3) opens the way to confirming/testing the mathematical hypotheses of existence of a mathematical theory T, because the model of T provides legitimate mathematical entities (sets of legitimate mathematical entities) of which the existence hypotheses of the theory are true;

(4) has a better agreement with the history of mathematics;

(5) receives an indirect confirmation from the observation that finding the model of a mathematical theory T within an already accepted mathematical theory T' is what often leads to the possibility of applying T to the empirical sciences.[31]

7 Some case-histories

Clearly, to be able to judge the correctness of the explanations given by the realist and the anti-realist of the philosophical relevance of the criterion of the model, I, once more, need to discuss paradigmatic cases in which such a criterion has been applied by the mathematical community.

In what follows, I shall examine three particularly important case-histories in which the criterion of the model performs a central rôle in the choice of theory operated by the mathematical community.

The first two case-histories are of theories that have been accepted by the mathematical community, whereas the third concerns a mathematical theory that has been rejected.

Theory introduction: the complex numbers

The complex numbers were introduced for the first time by the Italian mathematician Raffaello Bombelli (1572) in the discussion of a particular case of resolving Tartaglia's formula for cubic equations, namely, the case in which all the three roots are real and different from 0. In this case, in the relevant formula, there always occur square-roots of negative numbers.[32]

Complex numbers are extremely important in many branches of mathematics. They allow a solution for the equation $x^2 + 1 = 0$ (an equation which has no solution in $\mathbb{R}$), underpin many important theorems such as the fundamental theorem of algebra — concerning the existence of roots of a polynomial equation $p(x) = 0$ — provide extensions to analysis, etc.[33]

However, notwithstanding the obvious importance that complex numbers have in the fields of algebra, analysis, etc. they were not generally accepted before the interpretation given to them by Gauss according to which there

---

details on this point.)

[31] This is, indeed, the case, because, if T is applicable to the empirical sciences, we can then use an argument based on the Indispensability Thesis to add considerably more weight to the realist's explanation of the criterion of the model.

[32] [Boyer, 1968], Ch. XV, §13, p. 316.

[33] See [Tonelli,], vol. I, Cap. I, §3.15, pages 21/22.

exists a biunivocal correspondence between the set of complex numbers $\mathbb{C}$ and the set of points of the plane:[34]

> It could be said in all this that so long as imaginary quantities were still based on a fiction, they were not, so to say, fully accepted in mathematics but were regarded rather as something to be tolerated; they remained far from being given the same status as real quantities. There is no longer any justification for such discrimination now that the metaphysics of imaginary numbers has been put in a true light and that it has been shown that they have just as good a real objective meaning as the negative numbers.

This fact, which as Gauss himself says, is related to the existence and meaningfulness (representability) of complex numbers (points (1)—(3), §2.6), deserves careful consideration, because it is neither the mere manifestation of an historical event which entirely depends on the mathematical fashion of the time, nor is something confined to a particular mathematical theory. On the contrary, it shows a constant pattern present within mathematical activity according to which the existence and mathematical meaningfulness of a given set of however useful newly introduced entities can only be established through their interpretation given in terms of *respectable* and already accepted mathematical entities (point (4), §2.6).

Furthermore, as is well known, the representation of $\mathbb{C}$ in terms of ordered pairs of real numbers is at the root of the application of complex numbers to physics. In fact, if $z \equiv (a, b)$, for $z \in \mathbb{C}$ and $a, b \in \mathbb{R}$, (a, b) can be seen as what describes a vector in $\mathbb{R}^2$; $z + w$, for $z, w \in \mathbb{C}$, as the sum of two vectors, and $k \times z$, where $k \in \mathbb{R}$ and $z \in \mathbb{C}$, as the product of a scalar times a vector. And since we can interpret the concept of force applied to a body b moving on a plane α as a vector in $\mathbb{R}^2$, we can show the relevance of the complex numbers to mechanics *via* the model of $\mathbb{C}$ (point made at the end of §2.6). In actual fact,[35]

> What the complex numbers do for vectors in a plane is to supply an algebra to represent the vectors and operations with vectors. One need not carry out the operations geometrically but can work with them algebraically much as the equation of a curve can be used to represent and work with curves.

Theory introduction: the non-Euclidean geometries

Let us now consider the case-history represented by the non-Euclidean geometries.

Another very striking case of the acceptability of mathematical theories being related to the discovery of models is that of the introduction of non-Euclidean geometries within the domain of mathematical theories.

---

[34][Gauss, 1863], Vol. 10, 1, p. 404 in [Ebbinghaus *et alii*, 1991], Ch. 3, p. 62.
[35][Kline, 1972], Ch. 32, §. 2, p. 777.

7. SOME CASE-HISTORIES

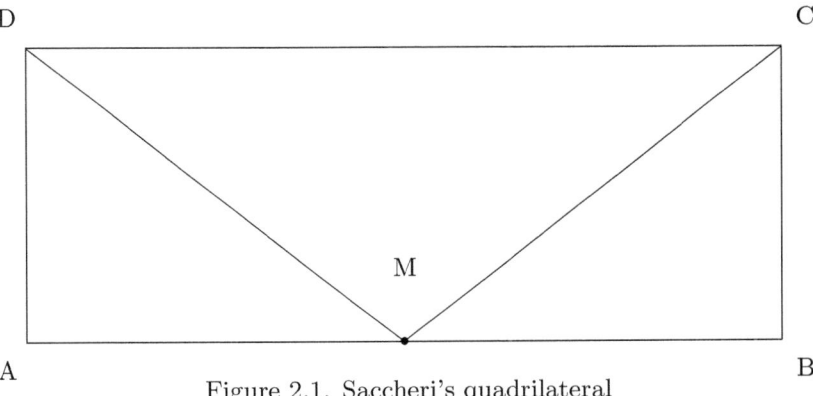

Figure 2.1. Saccheri's quadrilateral

If we call 'absolute geometry'[36] the set of theorems derivable from the subset of Euclid's axioms which does not include the Axiom of Parallels or any other proposition equivalent to it, we can obtain hyperbolic geometry by adding to the axioms of absolute geometry the *hypothesis of the acute angle*. Such an hypothesis says that 'If $ABCD$ is a Saccheri quadrilateral then $\angle ADC = \angle BCD < 90°$' (HAA). (The assertions 'If $ABCD$ is a Saccheri quadrilateral then $\angle ADC = \angle BCD = 90°$' and 'If $ABCD$ is a Saccheri quadrilateral then $\angle ADC = \angle BCD > 90°$' are respectively called 'the hypothesis of the right angle' (HRA) and 'the hypothesis of the obtuse angle' (HOA).)

A Saccheri quadrilateral is a four-sided figure $ABCD$ such that $\overline{AD} = \overline{BC}$, $\angle DAB = \angle CBA = 90°$:

Given a Saccheri quadrilateral, it is possible to prove within absolute geometry that $\angle ADC = \angle DCB$[37]. What is independent of absolute geometry is deciding whether $\angle ADC = \angle DCB = 90°$ or not.

In absolute geometry it is possible to prove that HRA is equivalent to the Axiom of Parallels and, therefore, adding HRA to absolute geometry would generate Euclidean geometry. However, simply adding HOA to absolute geometry would not generate any geometry at all for the two following

---

[36] I take the term 'absolute geometry' from [Kelly & Matthews, 1981], Ch. 1, §5, p. 17.

[37] If M is the middle point of $\overline{AB}$ and we conjoin C and D with M, we realise that $\triangle ADM \equiv \triangle MCB$, because by hypothesis $\overline{AD} = \overline{CB}$, $\angle DAM = \angle CBM$, and $\overline{AM} = \overline{MB}$ since M is the middle point of $\overline{AB}$. Therefore, by one of the congruence criteria for triangles we have that $\triangle ADM \equiv \triangle MCB$. From this we can derive two things: 1) $\angle MDA = \angle MCB$ and 2) $\overline{DM} = \overline{MC}$. Now 2) implies that $\triangle DMC$ is isosceles and that, therefore, $\angle MDC = \angle MCD$. Obviously, $\angle MDC = \angle MCD$ and $\angle MDA = \angle MCB$ imply that $\angle MDA + \angle MDC = \angle MCB + \angle MCD$, i.e. $\angle ADC = \angle BCD$.

reasons.

First, although in elliptic geometry lines are *boundless*, i.e., a line in the elliptic plane has no end points, they, nevertheless, have finite length, whereas in absolute geometry lines are assumed to have infinite length.[38]

Secondly, since the HOA is equivalent to the assertion that 'Two lines in a plane always intersect', we can deduce that in elliptic geometry there are no parallel lines. This fact is plainly inconsistent with the provable existence of parallel lines within absolute geometry.[39] Therefore, in order to obtain elliptic geometry from absolute geometry several changes must be made to the axiomatic basis of absolute geometry.

In the atmosphere of general disinterest that followed, with a few happy exceptions, the publication of Lobatchevsky's and J. Bolyai's systems of hyperbolic geometry, the cries of the relatively small number of 'Bœotians' who had become aware of these contributions were stopped in their throats by Beltrami who was the first to give a Euclidean model for H using the pseudo-sphere.[40] He, by interpreting an hyperbolic line as a geodesic on the pseudo-sphere, showed that the geometry of a limited region of the hyperbolic plane is the same as the geometry of part of the surface of the pseudo-sphere.[41]

In the course of time also Klein and Poincaré constructed Euclidean models for hyperbolic geometry by representing hyperbolic plane geometry in a Euclidean circle,[42] and it was shown that the surface of a sphere is a Euclidean model for elliptic plane geometry when the elliptic line is interpreted as a great circle on the sphere.[43]

As we shall see in the following quotation, the importance of Beltrami's, Klein's, and Poincaré's work which resulted in the construction of Euclidean models for non-Euclidean geometry is not confined simply to a proof of the relative consistency or to the clarification of the meaning of terms belonging to non-Euclidean geometry (point (1), §2.6), but also extends to providing a good way of representing hyperbolic lines, planes, and geometrical figures (points (2)–(3), §2.6):[44]

> Kant's proof of the *a priori* origin of the geometrical axioms is based on the asssertion that it is impossible to form a mental representation of space-relations at vari-

---

[38]See [Tuller, 1967] §1.6, p. 14. The assumption of the infinity of the line is necessary to prove Euclid's theorem of the exterior angle.

[39]See [Greenberg, 1980] Corollary 2 to theorem 4.1: 'If l is any line and P is any point not on l, there exists at least one line m through P parallel to l', Ch. 4, p. 95.

[40]See on this [Bonola, 1955], Ch. IV, pp. 84-128.

[41]See [Tuller, 1967], §2.7, page 27.

[42]For more details on Klein's and Poincaré's Euclidean models of hyperbolic geometry see M.J. Greenberg, Ibid., Ch. 7.

[43]See [Tuller, 1967], §1.6, p. 16.

[44][Helmholtz, 1878], in [Ewald, 1996], pp. 686-7.

7. SOME CASE-HISTORIES

ance with Euclid's geometry. But the 'metamathematical' investigations passed under review in my former paper have shown that it is quite possible to devise and consistently work out systems of geometry that differ from Euclid's both in the number of space-dimensions and in their axioms, with their related systems of mechanics ... Beltrami's discovery of the way of representing pseudospherical space in a sphere of Euclidian space shows directly what would be the appearance of optical images in pseudospherical or spherical space. Every optical image of objects at rest as seen by a spectator at rest would, in fact, be exactly the same as that of the corresponding representation in Beltrami's sphere as seen from the centre (supposing always that the distance of the two eyes may be neglected in comparison with the imaginary radius-of-curvature of the space). There would be a difference only in the order of succession of the images, according as the observer or the solid objects moved. Nothing would be changed but the rule for inferring what images would succeed others in case of movement. And, as I have maintained, such differences are not necessarily considerable, nor need they excite attention. Men lived for a long time on what they thought was the flat earth, before they discovered its spherical form, and they struggled long enough against this truth, just as our Kantians at the present day will not listen to the possibility of representing pseudospherical space.

At this point someone can say that, if it is true to assert that in the past a number of controversies concerning the acceptability of certain theories were brought to an end by the production of models of such theories given in terms of already accepted mathematical theories, nowadays[45]

> Since we are ... conceptually more sophisticated, this is not likely to happen in the same way again

It is to show the correctness of my claim as something related to a constant feature of the debate concerning the acceptability conditions for mathematical theories (point (4), §2.6) that I am going to discuss the case of Quine's system New Foundations (**NF**). This is a particularly interesting case, because, as a consequence of the lack of a model for **NF**, this system has not been accepted by the mathematical community.

Theory elimination: Quine's New Foundations

The system **NF** was devised by Quine to eliminate the difficulties which affect type theory systems such as that implemented in *Principia Mathematica*. **NF** is a first-order system with only one sort of individual variables $x, y, z, \ldots$, and one two-place predicate $\in$. Given the set of well-formed formulae (wffs) of **NF**, Quine distinguishes between the *stratified* wffs, and the non-stratified ones.

A wff ϕ of **NF** is stratified just in case we can assign types (superscripts) to its terms in such a way that for any occurrence of $\in$ in ϕ the term immediately preceeding $\in$ has a type which is the immediate predecessor of the type assigned to the term immediately following $\in$.

---

[45][Dummett, 1994], p. 306.

It is important to point out that, although we can assign to different terms of ϕ the same type, the operation of type-assignement must be carried out uniformly. Moreover, if in ϕ there is the occurrence of an abstraction term such as $\{x : P(x)\}$ the type to be assigned to it is higher by 1 than the type assigned to x.

For example, the following wff is stratified:

$$(a) \quad ((x \in y) \implies (y \in z))$$

because

$$((x^1 \in y^2) \implies (y^2 \in z^3)).$$

On the other hand, (b) is clearly not stratified, for $1 \leq n$,

$$(b) \quad (x \in y_1 \in \cdots \in y_n \in x)$$

because

$$(x^m \in y_1^{m+1} \in \cdots \in y_n^{m+n} \in x^m),$$

for any type m. In particular

$$(x^m \in x^m)$$

is not stratified.

Some of the main advantages of **NF** are that: (1) through the concept of stratification, it embodies the vicious circle principle getting rid of Russell's paradox ($\neg(x^m \in x^m)$ is not a stratified wff, for any $m \in \mathbb{N}$) without discarding the unstratified wffs (as in **RTT**), but preventing any harmful effect caused by them through an opportune Zermelo-like restriction of the Principle of Comprehension according to which

Principle 7.1 (NF-comprehension).

$$\exists y \forall x (x \in y \leftrightarrow \phi(x))$$

where $\phi(x)$ is a *stratified* formula in which y does not occur free;

(2) it is possible to prove within **NF** the existence of an infinite set dispensing in this way with the Axiom of Infinity.

However, the disadvantages are that: (i) it is not excluded that Cantor's paradox may be constructed in **NF**; (ii) the proof of Cantor's theorem cannot be carried out in **NF**; (iii) the Axiom of Choice is provably false in **NF**; and, worst of all, (iv) no model has yet been found of **NF**. This last problem affecting **NF**, far from simply casting doubts on the consistency of this system, makes of stratification just a linguistic device which provides a merely *ad hoc* solution to the problem of guarding against some of the known paradoxes and it is what has so far induced the mathematical community to reject Quine's system as a foundation for mathematics.

8 The faithful representations argument

According to the *faithful representations* argument, we are justified in believing in the existence of terms belonging to the language of mathematical theories whose reference is not empty, because such a belief provides a very simple explanation of the significance of certain results, and of some of their consequences.

An example of this phenomenon is given by the following expressions (1)–(3), because postulating the existence of an object described in different ways by (1)–(3) — a circle with radius r — provides an obvious way of unifying them:

1. The conic section obtained intersecting $\mathbb{R}^2$ with a cone C in such a way that: (i) $\mathbb{R}^2$ is perpendicular to the axis of C; (ii) the intersection of $\mathbb{R}^2$ with the axis of C is (a, b), where $a, b \in \mathbb{R}$; and (iii) the distance between (a, b) and any point belonging to the conic section is r;

2. The set of boundary points of the open disk in $\mathbb{R}^2$ with centre (a, b) and distance r;

3. The set of solutions of the equation $(x - a)^2 + (y - b)^2 = r^2$, for $x, y, a, b, r \in \mathbb{R}$.

The phenomenon I have just described prompts mathematicians to interpret certain assertions in terms of different faithful representations[46] which produce information about a particular object, information which, adequately integrated, seems to generate an accurate picture of the entity represented, in the same way as different techniques of microscopy produce different pictures — and information — concerning the object observed.

For the realist, this shift towards the postulation of the existence of mathematical entities cannot be dismissed as what 'presents the advantages of theft over honest toil', because its coming about cannot be entirely explained in terms of a desire for harmony which responds only to aesthetical considerations.

In fact, the realist argues, the correctness of postulating the existence of a mathematical object, which is represented in several different ways, is confirmed by the observation that this produces a very natural explanation of the non-simply additive increase in knowledge that we witness as a consequence of the interaction of several of what I have called 'faithful representations'.

---

[46] A *representation* of an object is a description of the object expressed by propositions. A given representation of an object is *faithful* just in case the proposition (or propositions) which express it is (are) true.

A typical example of this phenomenon is the following. If we have two points **p** and **q** in a plane α, we can give an analytical representation of these two points by means of ordered pairs of real numbers (x_1, y_1) and (x_2, y_2). And, we can also give a geometrical representation of the two points by saying that they are the extreme points of the segment $\overline{\mathbf{pq}}$.

Now, if we want to define the distance between **p** and **q**, we obtain an answer to this problem by means of the interaction of the two types of representations *via* a geometrical result — the Pythagorean Theorem — which was obtained for a totally different purpose from that of measuring the distance of any two points in a plane. Our answer is of the form:

$$d(\mathbf{p}, \mathbf{q}) = \sqrt{\mid x_1 - x_2 \mid^2 + \mid y_1 - y_2 \mid^2}.$$

To this argument the anti-realist can reply that the non-simply additive growth of information we have in mathematics whenever there is interaction between different mathematical theories is a phenomenon which is not common only to those theories that describe some kind of reality.

"In the case of cartoon characters", the anti-realist says, "we observe that we can 'represent' them either by means of descriptions given in language or by drawing them on paper; and that when we combine these two different ways of representing fictional characters in a cartoon, we produce a non-simply additive increase of information about them which is, of course, independent of the existence of entities to which we refer by means of terms like 'Charley Brown', 'Lucy', 'Snoopy', 'Donald Duck', etc."

However, the realist can avoid capitulating to the cartoon-objection observing that, in the case of cartoon characters, language (the dialogue) and drawing are not descriptive of entities which exist independently of the activity of the cartoonist, but are *both* constitutive of the cartoon characters. Consequently, the interaction between language and drawing in a cartoon is obvious, immediate, and *ad hoc*, because the drawing is deliberately planned to go with the dialogue.

On the other hand, the realist adds, this is not the case in mathematics. For, in mathematics the interactions between different mathematical theories are not set up or agreed upon. They are often discovered a long time after the introduction of the mathematical theories involved. An interesting example of this phenomenon may be found in Pythagorean mathematics.

According to the Pythagoreans, the natural numbers were configurations of points, like the elements of the sequence in fig. 2.2, which were by them called 'triangular numbers', or like the elements of the sequence in fig. 2.3, which they called 'square numbers'.

The history of mathematics tells us that it was a genuinely surprising discovery, and not the outcome of a conspiracy, to find out, sometime after

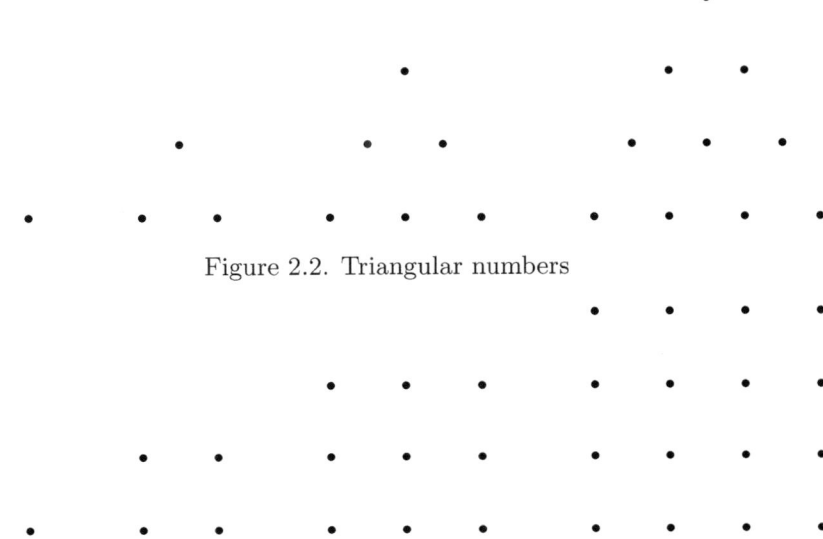

Figure 2.2. Triangular numbers

Figure 2.3. Square numbers

the introduction of the triangular and the square numbers, that a triangular number has the following arithmetical shape (representation):

$$1 + 2 + \cdots + n, \text{ for some } n \in \mathbb{N},$$

and that:

$$1 + 2 + \cdots + n = \frac{n(n+1)}{2}.$$

What I have called 'arithmetical representation' of triangular numbers allows us to prove arithmetically results about triangular and square numbers which were originally proved geometrically,[47] and increase, in a non-*ad hoc*, non-simply additive way, the information in our possession about triangular, square, etc. numbers.

9 The indispensability thesis

The applicability of mathematics to the empirical sciences has been an object of great interest for many philosophers. Frege, for example, arguing

---

[47] One such result is the following:

Theorem 8.1. The sum of any two consecutive triangular numbers is a perfect square.

against the formalists, asserted that:[48]

> ...it is applicability alone which elevates arithmetic from a game to the rank of a science.

But, of course, having made an assertion like the one above, it is crucial to spell out how arithmetic is applied, and whether the way it is applied implies any ontological commitment to the existence of numbers, etc.

For Frege:[49]

> The laws of number ... are not really applicable to external things; they are not laws of nature. They are, however, applicable to judgements holding good of things in the external world: they are laws of the laws of nature. They assert not connexions between phenomena, but connexions between judgements; and among judgements are included the laws of nature.

However, in his reflections on the applicability of arithmetic, Frege does not seem to consider the possibility that the applicability of the laws of number to the laws of nature might be purely accidental, in the sense that there could be ways of formulating the laws of nature which would dispense with the laws of number.

Indeed, if this were the case then Frege could not justify his claim that the laws of number are the laws of the laws of nature, and consequently the formalists might have a point in claiming that, after all, applicability does not elevate arithmetic from a game to the rank of a science. It is to close this gap in the reasoning from the applicability of arithmetic to realism about arithmetic that the Indispensability Thesis has been formulated and defended.

One of the most compact expositions of the Indispensability Thesis, and of the argument in favour of mathematical realism based on it, is that given by P. Maddy in what follows:[50]

> Our best scientific theory of the world makes indispensable use of mathematical things: the temperature of a gas as a function of time, acceleration as a second derivative, Maxwell's equations. (This is taken as an unvarnished fact.) To draw a testable consequence from our theory requires the use of various far-flung parts of that theory, including much mathematics, so the confirmation resulting from a successful test adheres not to individual statements but to large bodies of theory. (This is holism.) Finally, our theory is committed to those things that it says 'there are'. (This is Quine's criterion of ontological commitment.) It follows that our theory, and we who adopt it, are committed to the existence of mathematical things.

---

[48] [Frege, 1977], §91, p. 187.
[49] See [Frege, 1884], §87, p. 99.
[50] [Maddy, 2000], Part II, Ch. 6, p. 133.

10. FIELD AND THE INDISPENSABILITY THESIS

The main objections against the argument above can be divided into two classes: (i) those against the very idea that 'our best scientific theory of the world makes indispensable use of mathematical things'; and (ii) those which accept that mathematics is indispensable to our best theory of the world, but challenge the conclusion that we should have an ontological commitment to the posits of the mathematical theories which are applied.

In the critical assessment of the argument from indispensability that I am going to develop in the remaining part of this chapter I will, first, consider an important objection of type (i): H. Field's view of the relation existing between mathematics and the empirical sciences; and I will then discuss some interesting examples of type (ii) objections.

10 Field's argument against the Indispensability Thesis

For Field:[51]

> ...to explain even very complex applications of mathematics to the physical world ...it is not necessary to assume that the mathematics that is applied is true, it is necessary to assume little more than that mathematics is consistent.

He intends to attack the indispensability thesis showing that:[52]

> ...the mathematics needed for application to the physical world does not include anything which even *prima facie* contains references to (or quantifications over) abstract entities like numbers, functions, or sets. Towards that part of mathematics which does contain references to (or quantifications over) abstract entities...[Field says] I adopt a fictionalist attitude: that is, I see no reason to regard this part of mathematics as *true*.

Field argues this point defending the claim that:[53]

> ...one can always reaxiomatize scientific theories so that there is no reference to or quantification over [abstract] mathematical entities in the reaxiomatization (and [that] one can do this in such a way that the resulting axiomatization is fairly simple and attractive.)

This is in principle possible for him, because, according to Field:[54]

> ...if you take any body of nominalistically stated assertions N,[55] and supplement it with a mathematical theory S, you don't get any nominalistically-stable con-

---

[51][Field, 1980], Preface, p. vii.
[52][Field, 1980], Preliminary Remarks, pp. 1-2.
[53][Field, 1980], p. viii.
[54][Field, 1980], Chapter 1, p. 9.
[55][Field, 1980], footnote 8, p. 108:

> The formal content of saying that N is 'nominalistically statable' is simply that it [does] not overlap in non-logical vocabulary with the mathematical theory to be introduced.

clusions that you wouldn't get from N alone. [This is more rigorously expressed by Principle C, p. 12.]

It is important to observe that Field has no qualms about the methods which should be adopted to prove that 'abstract entities are not needed for ordinary inferences about the physical world or for science'. For, he says,[56]

> ...if I am successful in proving *platonistically* that abstract entities are not needed for ordinary inferences about the physical world or for science, then anyone who wants to *argue* for platonism will be unable to rely on the Quinean argument that the existence of abstract entities is an indispensable assumption.

He then begins to carry out his project through an attempt to give a nominalistic account of the structure of physical space. This is done in two steps.

First, Field considers that, in a theory broader than Euclidean geometry (EG), Hilbert proved a representation theorem of EG over the reals and other theorems establishing the uniqueness of such a representation. For Field, the above mentioned theorems are very important. In fact, on the one hand,[57]

> Hilbert's representation theorem ...shows that statements that talk about space alone, without reference to numbers, are equivalent to certain 'abstract counterparts' which do talk about numbers. Because of this, we can use the theorem as a device for drawing conclusions about space (conclusions *statable without* real numbers) much more easily than we could draw them by a direct proof from Hilbert's axioms ...So invoking real numbers (plus a bit of set theory) allows us to make inferences among claims not mentioning real numbers much more quickly than we could make those inferences without invoking the reals. And the inferences we make in this way will be correct every time.

and, on the other hand, Hilbert's uniqueness theorems appear to support the idea that:[58]

> ...the fact that geometric laws, when formulated in terms of distance, are invariant under multiplication of all distances by a positive constant, but are not invariant under any other transformation of scale, receives a satisfying *explanation*: it is explained by the *intrinsic facts* about physical space, i.e. by the facts about physical space which are laid down without reference to numbers in Hilbert's axioms.

The second step taken by Field in his attempt to give a nominalistic account of the structure of physical space consists in the realization that, if[59]

---

[56] [Field, 1980], pp. 5-6.
[57] [Field, 1980], Chapter 3, pp. 27-29.
[58] [Field, 1980], p. 27.
[59] [Field, 1980], Chapter 5, §II, pp. 37-38.

11. CRITICAL REMARKS   65

> ...we can regard the second-order quantifiers in Hilbert's theory as ranging over regions ... [and] write Hilbert's theory in this way, then the quantifiers (both first-order and second-order) range *only* over regions of space; and I've argued that regions of space are nominalistically acceptable entities.[60] So *if we write Hilbert's formulation of the Euclidean theory of space in this way, it has a purely nominalistic ontology.*

On the basis of what he has done to give a nominalistic account of the structure of physical space, Field attempts to generalize his approach extending it to space-time.

According to Field[61]

> What we would like is to do for space-time what has been done for space. That is, we want to come up with a system of 'intrinsic' axioms, more or less analogous to Hilbert's but involving somewhat different concepts, and to come up with a representation theorem that explains the legitimacy of coordinatizing space-time and a uniqueness theorem that explains why in the coordinatized treatment of space-time the laws of Newtonian mechanics will be invariant under just the coordinate transformations that they are in fact invariant under.

At this point Field turns his attention to sketching a possible nominalistic treatment of quantities (scalars) along the following lines:[62]

> A possible approach to a coordinate-independent treatment of, say, temperature, would be to introduce a continuum of temperature properties, each one the property of having such and such specific temperature. One could then describe the structure of that system of properties not via numbers, but via certain intrinsic relations among them, say the relations of betweenness and congruence; and one could impose axioms on these notions to guarantee that there was a $1-1$ function mapping the temperature properties into the reals, and that such a function was unique up to linear transformation.

He then goes on to give a nominalistic account of continuity, etc. as a further step towards producing a nominalistic version of Newtonian gravitational theory.

11 Critical remarks

Several are the objections which can be found in the literature against Field's argument concerning the indispensability thesis. I shall here confine myself to mentioning only some of those I take to be the strongest, and then expound five new ones.

---

[60][Field, 1980], p. 37.

> ...it seems attractive to regard points of space-time as a special case of regions, namely as regions of minimal size. So it seems to me that regions are nominalistically acceptable.

[61][Field, 1980], Ch. 6, p. 51.
[62][Field, 1980], Ch. 7, p. 55.

According to David Malament, the second-order language L within which Field's nominalist reformulation of physical theories[63] must be carried out is both too weak for physics and too rich for nominalism.

L is too weak for physics, because there are assertions concerning physical theories which cannot be re-expressed within L.[64] And it is too rich for nominalism, because: (a) a nominalist is not entitled to use second-order logic;[65] and (b) it is inadmissible for a nominalist to quantify over space-time points or space-time regions.[66]

A possible reply to the objection that L is too weak for physics is that, for this argument to cut any ice against Field's programme for the nomi-

---

[63][Malament, 1982], p. 526:

Let L be a second-order language with variables for individuals, variables for sets of individuals, and the relation symbols: '=', 'Seg-Cong', 'Scale-Bet', 'Scale-Cong', '$\in$', '$\leq$' (all of the appropriate type). Let a *standard interpretation* of L be one in which '$\in$' is assigned the (real) membership relation.

[64][Malament, 1982], p. 529:

Field's nominalist physicist cannot assert that it is possible for the Klein-Gordon field to be nonconstant. He cannot assert that the field evolves deterministically. Indeed he cannot do anything except assert general truths about what goes on within arbitrary models.

[65][Malament, 1982], p. 530:

Presumably, if a "logic" is not recursively axiomatizable (and not compact) its logical-consequence relation cannot be recovered in terms of a formal derivation system. So, the objection goes, a nominalist cannot understand the assertion that sentence S_L is a logical consequence of theory T. (What could it mean to say that S_L is true in all set-theoretic models of T?)

In connection with this objection, consider again the example used to illustrate Field's remarks about conservativeness. The claim there was that the detour derivation of S_L from T making use of auxiliary mathematical hypotheses is justified insofar as those hypotheses are conservative. For then, according to Field, one can always prove S_L from T alone. But how, in the absence of a formal derivation system, is a nominalist supposed to do *that* (in general)?

[66][Malament, 1982], p. 532:

Suppose it is agreed that space-time points are "entities that exist in their own right." Still, philosophers with nominalist scruples might well be uncomfortable with them. They certainly are not concrete physical objects in any straight-forward sense. They do not have mass-energy content (unlike, for example, the Klein-Gordon field itself). They do not suffer change. It is not even clear in what sense they exist *in* space and time.

Field takes for granted the distinction between concrete, physical objects on the one hand and abstract objects on the other. But I, for one, begin to lose my grip on the distinction when thinking about such things as "space-time points." It would have helped me to understand his conception of nominalism if Field had explained how he draws the line and made clear why space-time points are so much *better* than, for example, sets and qualities.

11. CRITICAL REMARKS

nalization of physics, Malament, besides producing assertions belonging to physical theories which cannot be re-expressed within L, must also show that such assertions have great physical significance, and that they cannot even be 'paraphrased' within L.

Concerning Malament's point on L being too rich for nominalism, this has received a very large consensus among the critics of Field, a consensus which has been in part motivated by some of the reasons put forward by Malament. For example, the observation that the nominalist is not entitled to use second-order logic is, on the one hand, strongly justified in the literature by what Malament himself says on this (see footnote 65) and, on the other hand, by an appeal to the well known reasons behind the famous Quinean quip: second-order logic is set theory in sheep's clothing.[67]

However, it seems to me that the situation is not so clear cut when it comes to assessing Malament's claim that it is inadmissible for a nominalist to quantify over space-time points and space-time regions. For, if it is the case that: space-time points do not have mass-energy content, we do not have identity criteria for them, we cannot think of an experiment which reveals to us the existence of a space-time point, etc. it is also the case that there exists experimental evidence to the effect that some of the properties of the distance function (the metric) between points of space-time are affected by the mass of the bodies present in space-time.

In any case, independently of the considerations above, concerning whether it is admissible for a nominalist to quantify over space-time points and space-time regions, we must observe that, if P. Maddy is right when she says that '...the question of the continuity of space-time must be considered open',[68] because a number of 'difficulties in theories ranging from classical electrodynamics to quantum gravity have been traced to the assumption that space-time is continuous,'[69] it follows that Field cannot simply assume that the physical world contains *traditional* space-time points and space-time regions.[70] This is a very serious problem for Field, because

---

[67] Even though several attempts have been made to deflate the ontological commitment of second-order logic through, for instance, the concept of 'plural quantification' (see on this [Martino, 2004]), the 'Quinean worries' about the ontological commitment of second-order logic have not been put to rest.

[68] [Maddy, 2000], Part II, Ch. 6, §(ii), p. 151.

[69] [Maddy, 2000], Part II, Ch. 6, §(ii), p. 151.

[70] Field holds a substantival view of space-time, that is, for him ([Field, 1989a], pp. 171–172):

> ...the physical world contains not only aggregates of matter (physical objects, their spatio-temporal parts etc.), but also (over and above these, i.e., not logically constructed from them) space-time and its spatio-temporal parts. A 'part of space-time' is of course just a space-time region.

the impossibility of assuming without justification that the physical world contains *traditional* space-time points and space-time regions undermines the foundations of his entire nominalization programme.

In his review of *Science without numbers*, Malament's last complaint against Field's nominalization programme, a complaint with which I entirely sympathize, is that the examples used by Field to show how his nominalization programme ought to be carried out are not representative, because there are physical theories the nominalistic reformulations of which require us either to quantify over possible dynamical states (classical Hamiltonian mechanics) or appeal to propositions (or eventualities) (quantum mechanics). And, clearly, as a consequence of their abstract nature, neither possible dynamical states nor propositions (nor eventualities) are admissible for a nominalist.

Resnik has two objections against Field's attempt '...to demonstrate that we can expunge mathematics from science by replacing analytic, mathematized scientific theories by synthetic versions that do not refer to mathematical objects,'[71] objections which differ from those examined so far.

The first consists in a very general claim to the effect that, according to expert opinion, there is '...little evidence that Field's approach can even get started with the job of reformulating quantum mechanics or general relativity theory.'[72]

The second is that, when we consider the application of synthetic theories to experimental data, above all in sciences others than physics, we find that a generous use is made of statistics and of statistical inference, that is, the use of statistics aimed at deciding whether '...a set of examples falsifies a statistical claim.'[73] And, of course, if this is the case, how can we hope to do science without numbers? (According to Resnik, a similar problem arises for Field with probability and probabilistic inference.)

In attempting to evaluate Resnik's objections, it must be said that the first is not compelling, because it consists in a not better specified appeal to authority.

Concerning the second objection, I think that, independently of the very ambitious title of Field's book, if Field's nominalization programme could be made to work in physics — so far the most mathematical of the empirical sciences — that of itself would be a very important result. In examining Resnik's second objection I shall, therefore, restrict my considerations to the narrower claim that it is possible to do physics without numbers.

Now, when we talk about the rôle played by statistics and probability

---

[71][Resnik, 2001], Ch. 4, §1, p. 55.
[72][Resnik, 2001], Ch. 4, §1, p. 56.
[73][Resnik, 2001], Ch. 4, §1, p. 57.

11. CRITICAL REMARKS 69

in physics, we must distinguish between an internal and an external rôle. (From here on, instead of repeating 'statistics and probability', I will simply say 'probability'.) I call 'external' the probability in physics which involves only the heuristics and the confirmation techniques of a particular theory; and 'internal', instead, the use of probability in certain physical theories, e.g., in quantum mechanics, where the physical law itself is essentially probabilistic.

If we consider the distinction above, we realize that: (α) what Resnik has in mind in formulating his second objection is the external use of probability, and that (β) the external use of probability is of no concern for Field. For, what he wants to nominalize are physical theories, and not the techniques which are used to discover new particles, etc.

However, as we shall see in the discussion of one of the last five objections (the third) that I am going to examine in what follows, the internal use of probability in physics is indeed fatal to Field's project.

According to the first of these objections, if Field's argument is aimed as an attack against Platonism in mathematics, what is the point of showing that we can have a science without numbers, if even the present day Platonist rejects the position known as 'Platonism about objects/numbers' in favour of Platonism about structures?

In fact, also in case Field's argument were to go through, it would say nothing against Platonism about structures, quite the opposite! For, the only thing shown by his main argumentative devices — the representation and uniqueness theorems — is the existence of a relation of isomorphism between different systems. But this result, rather than scoring a victory for nominalism over Platonism in mathematics, provides identity criteria for an abstract entity: the structure.

At this point someone might say 'I don't see why you take Platonism about structures to be untouched by Field's argument. If Field is successful, he shows that there is no need for structures "as abstract entities", it can all be done with the space-time structures he invokes'.

To this the Platonist about structures can reply that the issue raised in the counter-objection above is based on a conflation of the notions of *system* and *structure*,[74] and that, consequently, if Field were successful, the only

---

[74] What I mean by 'system' and 'structure' is expressed by the following quotation (in [Shapiro, 1997], Ch. 3, §1, pp. 73–74):

> ...a *system* [is] a collection of objects with certain relations. An extended family is a system of people with blood and marital relationships, a chess configuration is a system of pieces under spatial and "possible-move" relationships, a symphony is a system of tones under temporal and harmonic relationships, and a baseball defense is a collection of people with on-field spatial and "defensive role" relations. A *structure* is the abstract form of a system, highlighting the interrelationships

thing shown by him would be the possibility of dispensing with an abstract system, but not with a structure.

Indeed, for the Platonist about structures, the relation of isomorphism between systems provides criteria of identity for structures whose *ante rem* existence is the only plausible explanation of the very existence of isomorphisms between systems.

Therefore, if Field is going to convince the Platonist about structures of the correctness of the nominalistic stand on science, he needs to do more than simply appeal to the existence of isomorphisms between concrete (spatio-temporal) and abstract systems. He should also argue that structures exist only *in re*.

Second objection: Field's application of his *structuralist* approach to nominalize (scalars like) temperature raises a few questions: what is the advantage of *postulating* a continuum of temperature properties over believing in the real numbers? How is he going to justify this move? His reasoning is as follows: if we had a set of physical properties **T** isomorphic with $\mathbb{R}$, we could then (1) dispense with $\mathbb{R}$, getting rid of abstract entities, and (2) hold on to **T**, the elements of which are perfectly acceptable for a nominalist. But, the fundamental question which remains unanswered by Field is: how does he know that there is such a thing as **T**?

For, a Fregean type of Platonist about numbers, who believes that arithmetic and analysis do not apply directly to reality, believes that numbers are abstract objects and that they are indispensable to the theories belonging to the empirical sciences, not because he believes in the existence of isomorphisms between abstract numerical structures and structures whose domain is made out of concrete objects, but because, for this kind of Platonist, arithmetic and analysis only assert laws regulating connections between judgments, laws which, in particular, apply to the judgments used in the empirical sciences.

Therefore, if Field intends to argue against the Fregean Platonist, he has to show that, in the particular case offered by temperature properties, (a_1) **T** exists, (a_2) there exists an isomorphism between **T** and $\mathbb{R}$. But, in *Science without numbers*, Field does nothing of the sort.

Furthermore, assuming that such thing as **T** exists, the use of the relations of congruence and betweenness defined on **T** could give to Field the possibility of determining/expressing the relation $x \leq y$. But what he could not express/determine without numbers is, in case $x < y$, *how much* bigger y is than x. And this last is a problem of great physical significance.

---

among the objects, and ignoring any features of them that do not affect how they relate to other objects in the system.

11. CRITICAL REMARKS

To see this consider the second law of Gay-Lussac.

11.1 (Second law of Gay-Lussac). If we keep constant the volume of a certain amount of gas, the pressure of this amount of gas varies, as the temperature t of the gas varies, according to the following equation

$$p_t = p_0(1 + \alpha t),$$

where p_0 is the pressure of the gas at 0 degrees centigrade, p_t is the pressure of the gas at t degrees centigrade, and α is a constant which is the same for all gases ($\alpha = \frac{1}{273}$).

From the second law of Gay-Lussac, we can predict that: if we keep constant the volume of a certain amount of gas, and increase the temperature t of the gas by 1 degree centigrade, the pressure of the gas will increase by $\frac{1}{273}p_0$. Now, the difference of 1 degree centigrade between t and $t+1$ is of great physical significance, because on it depends the possibility of confirming or falsifying experimentally the second law of Gay-Lussac.

These considerations, besides revealing the importance in physics of determining how much a given magnitude is greater than another, are also very important with regard to Field's project, because they reveal that, if Field's project of nominalizing physics makes impossible to determine, like in the case of temperature properties, how much a given magnitude y is greater than another magnitude x, this turns out to cause a serious and unacceptable loss of information within the nominalized system.

Third objection: the theory of real numbers is indispensable in quantum mechanics, because probability plays an internal rôle in the laws of quantum mechanics and you cannot have probability without real numbers.

To see that probability plays an internal rôle in the laws of quantum mechanics consider that, according to Heisenberg's Uncertainty Principle, if we have a particle moving along a straight line, it is not possible to measure simultaneously, as accurately as we want, both the particle's position x and its momentum p_x. The best we can do is to determine such values with uncertainty intervals Δx and Δp_x, which are related to each other in the following way:[75]

$$\Delta x \Delta p_x \geq \frac{h}{2\pi}. \qquad (2.1)$$

Now, since

$$\frac{h}{2\pi} > 0,$$

---

[75] In (1), (2) and (3) h is Planck's constant.

the uncertainty intervals will be both greater than 0 and, therefore, the determination of position and momentum of the particle will have to be essentially probabilistic, and the accuracy in the measurement of the position of the particle will be inversely proportional to that of its momentum.

To this objection Field can reply that he thinks that 'probability can be done nominalistically, by invoking — as in the case of the nominalization of temperature — comparative probability predicates'. But, for comparative probability predicates like 'x is more probable than y' to be meaningful, we must know what the expression 'probability of x' means. And, of course, given the present day disagreement over what we should understand by the expression 'probability of x' we can either use Kolmogorov's axioms as an implicit definition of 'probability of x', where x is a variable ranging over events, or we can consider 'probability of x' as what expresses a primitive notion and replace the question 'How do we calculate the probability of x?' for the question 'What is the meaning of "probability of x"?'. Clearly, in either case, the real numbers turn out to be an indispensable ingredient of the answer.

Fourth objection: if the theory of real numbers is indispensable in quantum mechanics, and classical mechanics is in principle reducible to quantum mechanics, it follows that, in contrast with Field's argument, the theory of real numbers is indispensable in classical mechanics as well.

To argue for the reducibility of classical mechanics to quantum mechanics we must consider that since $p_x = m \cdot v_x$, we can divide both members of (2.1) by m obtaining (2.2)

$$\Delta x \Delta v_x \geq \frac{h}{2\pi m}. \qquad (2.2)$$

If m is the mass of an electron or of a proton, m will be very small and, consequently,

$$\frac{h}{2\pi m} \qquad (2.3)$$

will be very large. In this case there will be serious difficulties in measuring simultaneously, as accurately as we want, both x and v_x. But when m is the mass of a macroscopic object then (2.3) becomes very close to 0, and we can measure simultaneously both x and v_x as accurately as we want. From this follows that when quantum mechanics is applied to macroscopic objects it gives the same results as classical mechanics.

Fifth objection: in physics there are quantities of great physical significance. What this means is that, if these quantities were different from what they actually are, also our universe would be different from what it is. Such quantities of great physical significance are the so-called physical constants, e.g., c (the speed of light), h (Planck's constant), G (the gravitational constant), etc.

The importance of some of the physical constants is also connected with the fact that they happen to characterize momentous changes in the history of physics: c is one of the important and distinctive features of Einstein's theory of relativity, h is at the heart of the conceptual turn brought about by quantum mechanics, etc.

But, now, if we think over these matters in relation to Field's attempt to nominalize physics, we cannot help concluding that it is not at all clear whether it is possible to re-axiomatize the physical theories of which physical constants are part in such a way that 'there is no reference to or quantification over' numbers. This problem becomes particularly acute when we consider the so-called dimensionless constants.

We say that a physical constant has *dimensions*, if the numerical value of the constant depends on the units of measure used to express it. From this definition follows that c, h and G have dimensions.

It is important to notice that constants which have dimensions seem to have some kind of 'geometrical' feature in the sense that they behave very much like segments of a line whose length's numerical value changes as the unit of length chosen changes.

However, not all physical constants have dimensions. For example, the constant of fine structure a

$$a = \frac{e^2}{\hbar c} \quad (2.4)$$

$$= \frac{1}{137}, \quad (2.5)$$

where e is the electrical charge of the electron and $\hbar = \frac{h}{2\pi}$, is dimensionless. It is simply expressed by the rational number $\frac{1}{137}$.

The case represented by the dimensionless physical constants is even more intractable for Field's nominalization programme than that of the constants which have dimensions. In fact, in the case of the dimensionless constants even the loosely 'geometrical' aspect common to their dimensional counterparts is lacking. How can Field nominalize that?

12 More objections against the Indispensability Thesis

Having completed the discussion of Field's programme which, as we have seen, aims at showing that our best scientific theory of the world does *not* make indispensable use of mathematical things, time has now come to examine some examples of type (ii) objections against the Indispensability Thesis.

These, as remarked in §2.9, are objections to the Indispensability Thesis which accept that mathematics is indispensable to our best theory of the world, but reject the idea that we have an ontological commitment to the posits of the mathematical theories which are indispensable to our best theory of the world.

The miserliness objection

For those who believe in the miserliness objection, natural scientists tend to 'economize' on the adoption of assumptions, whereas this is not the case with set theorists who, for instance, reject the axiom $V = L$ (the universe of sets V coincides with the universe of constructible sets L), because of its miserliness.

This type of disanalogy in the way in which mathematics and physics are done is particularly important for our discussion. For, the supposedly non-miserly attitude of mathematicians towards the existence assumptions belonging to their theories might lead them into introducing entities which are not really needed for the deployment of the mathematical theories themselves and, *a fortiori*, for the deployment of the scientific theories to which the mathematical theories are indispensable. In this case, the success of a scientific theory could not be taken as a confirmation of the mathematical existence assumptions belonging to the mathematical theory which is indispensable to it.

There are various reasons why the miserliness objection does not work. In what follows I will mention only three of them.

First, the idea that natural scientists tend to 'economize' on their assumptions is far too vague to be of any use in this context. Were Archimedes, Galileo, Newton, Einstein, and countless other innovative thinkers, 'economizing' when they gave their daring contributions to physics?

Secondly, 'economizing', or applying Ockham's razor, is a current, and constant feature of mathematical activity. It is well known that in mathematics, given a set of axioms for a theory T, it is an important task to determine which is the smallest set of independent axioms capable of axiomatizing T.

Thirdly, what seems to have been overlooked by the propounder of the miserliness objection is that someone who believes in the correctness of the Indispensability Thesis can reply to him saying that the Indispensability Thesis makes a strong case for having an ontological commitment *only* to the posits of applicable mathematics.

12. MORE AGAINST THE INDISPENSABILITY THESIS 75

Therefore, any critical reference, on the part of the propounder of the miserliness objection, to a connection between the Indispensability Thesis and an ontological commitment to the posits of those areas of mathematics which are not applicable is out of place.

The double-standards objection

According to what I call the 'double-standards objection,'[76]

> ... mathematical existence assumptions in science, and their accompanying assumptions about the structure of physical reality, are not treated on an epistemic par with ordinary physical assumptions: the standards for their introduction are weaker, and their role in successful theory lacks confirmatory force; they are at once favoured and trivialized.

One of the consequences of this observation seems to be that, once more, the indispensability of mathematics to the deployment of the natural sciences does not imply a commitment into believing in the posits of the mathematical theories which are applied.

The objection above, like the miserliness objection, does not seem to be compelling. Indeed, the fact that physicists do not treat the mathematical existence assumptions they use in their work as seriously as the physical existence assumptions only shows that they are *not* doing mathematics.

On the other hand, mathematical existence assumptions within mathematics are treated very seriously indeed. The constant attempt to provide models for new mathematical theories in terms of theories which have already been accepted, and the correlative sceptical attitude manifested by the mathematical community towards theories for which models have not yet been found, speaks precisely in favour of this (see §7.3).

Moreover, one could readily agree with the idea that it is unreasonable to expect that mathematical existence assumptions in science be treated on an epistemic par with ordinary physical assumptions, because mathematical existence assumptions are about entities which do not occupy portions of space-time, whereas ordinary physical existence assumptions deal with concrete objects; and still hold that a successful scientific theory has confirmatory force concerning mathematical existence assumptions, because a successful scientific theory provides confirmation of the verisimilitude of the laws of nature expressed by it, and the verisimilitude of the laws of nature expressed by the successful scientific theory provides confirmation of the verisimilitude of the mathematical laws of these laws of nature and, in particular, of their existence assumptions.

Furthermore, a Quinean philosopher might say that, if we consider carefully enough the status of applicable mathematics and its existence assump-

---

[76][Maddy, 2000], ibid., §(iv), p. 156.

tions, and compare it with that of a physical theory and its existence assumptions, we realize that there is, after all, a way in which the existence assumptions of applicable mathematics, and those of physics, can be treated on an epistemic par.

In fact, if we are prepared to take seriously the phenomenon that Quine calls 'inscrutability of reference', we cannot but conclude that, even in the case of a scientific theory T, what matters to explaining the predictive value of T is the structure of T, and not the objects which might be in the domain of T.[77]

But, if this structuralist view of empirical science goes through, then there is a strong case for considering also physical objects, and therefore existence assumptions relating to physical objects, as (1) components of theories whose 'rôle in successful theory lacks confirmatory force'; and as (2) assumptions whose introduction into scientific theories *ought to be* 'at once favoured and trivialized'.

The idealization objection

'If the application of a mathematical concept, such as that of continuity, to physics is simply the outcome of an idealization', so the idealization objection goes, 'it follows that its indispensability to our best theory of the world cannot commit us to believing in the existence of the continuum'.

In evaluating the idealization objection, it seems that this, if correct, is certainly not applicable to the concept of continuity.

In fact, since, as a consequence of the Complementarity Principle, the wave-interpretation of electro-magnetic phenomena in quantum mechanics cannot be discounted, and since the concept of wave entails that of continuity, it follows that it cannot be discounted that continuity in physics, rather than simply being an idealization, is actually realized by electro-magnetic phenomena.

But, of course, if we cannot discount that continuity is realized by electro-

---

[77][Quine, 1981a], p. 20:

Structure is what matters to a theory, and not the choice of its objects. F. P. Ramsey urged this point fifty years ago, arguing along other lines, and in a vague way it had been a persistent theme also in Russell's *Analysis of Matter*. But Ramsey and Russell were talking only of what they called theoretical objects, as opposed to observable objects.

I extend the doctrine to objects generally, for I see all objects as theoretical.

and (Ibid., p. 22)

Our overall scientific theory demands of the world only that it be so structured as to assure the sequences of stimulation that our theory gives us to expect. More concrete demands are empty ...

12. MORE AGAINST THE INDISPENSABILITY THESIS

magnetic phenomena, it follows that the idealization objection is not applicable to the concept of continuity, because its premise — the application of continuity to physics is simply an idealization — cannot be shown to be true.

Secondly, if the idea that the concept of continuity is simply an idealization and is not realized by physical phenomena derives from the anomalies generated in classical quantum field theory by the conception of a space-time continuum associated with the idea that elementary particles are point-like, or have no internal structure, we must reply that there are theories, the so-called 'super-string theories', in which the two above mentioned ideas coexist without causing any anomaly.

Thirdly, and most importantly, a Fregean philosopher can say that the idealization objection is wide of the mark, because, since mathematics does not apply to external things but to the laws of nature, it is unreasonable to expect that what is postulated by its existence assumptions be realized by physical phenomena.

The limited commitment objection

For the supporters of the limited commitment objection, the Indispensability Thesis has a very limited value in the controversy between realists and anti-realists in the philosophy of mathematics, because it does not justify a complete commitment to believing in the existence of the posits of those mathematical theories which are indispensable to our best theory of the world.

'In fact', they argue, 'if we are holist with regard to scientific theories, the indispensability of some mathematical theories to our best theory of the world commits one, at the very most, to believing in the existence of the posits of such theories.

But, since, as the history of science shows, the range of mathematical theories which are indispensable to our best theory of the world changes over time, it follows that an argument from indispensability can only be considered as conveying, exclusively in *local* realism/anti-realism disputes and for a given time, that conviction typical of empirical confirmation, a conviction which speaks of a limited ontological commitment to the posits of a mathematical theory'.

The argument just offered by the believer in the limited commitment objection is not very convincing. For, given that mathematical theories are conceivable as sub-theories of set theory, the indispensability of some of them to our best theory of the world forces us to have an ontological commitment to the posits of set theory regardless whether or not the mathematical theories which are indispensable to our best theory of the world change over time. And, if we believe that all there is to mathematics is set theory (plus first-order logic), an ontological commitment to the posits of set theory would, indeed, resolve the issue of 'global realism' about mathematics.

At this point the defender of the limited commitment objection could say that, even if we accept the idea that there is nothing more to mathematics than set theory (plus first-order logic), we must recognize that some of the existence assumptions/posits of standard set theory can be used to prove the existence of sets of such a huge size that they are neither applied to our best theory of the world nor is it presently conceivable how they could become indispensable to it in the foreseeable future.

From this follows that, since not all the existence assumptions/posits of set theory are necessary to derive the sub-theories of set theory that are indispensable to our best theory of the world, arguments from indispensability turn out to be compelling only for those posits of set theory which are necessary for the derivation of those sub-theories of set theory which are actually indispensable to our best theory of the world.

In spite of its immediate appeal, the objection above is not compelling, because one can reply that: (1) set theory comes as a whole, and unless we are able to draw a meaningful demarcation line between the part of it (and its posits) which is indispensable to our best theory of the world and that which is not, preserving the mathematical dignity of the theory, we have a commitment towards *all* the posits of set theory; and that (2), given how the mathematical theories which are indispensable to our best theory of the world change over time, it is very doubtful that such a demarcation line can ever be drawn.

In any case, the supporter of the relevance of the arguments from indispensability to global realism concerning mathematics could always either gerrymander, however unnaturally, the grammar of set theory so to exclude its higher reaches or develop systems of predicative set theory or could, more simply, 'bite the bullet' and call 'set theory proper' only that shrunken part of current set theory which has as existence assumptions/posits only those necessary to derive the mathematical theories which are indispensable to our best theory of the world, and consider the rest of set theory as ideal statements in Hilbert's sense.

CHAPTER 3

TRADITIONAL REALIST POSITIONS

1 Introduction

To complete the review of the case made in favour of realism in mathematics (set-theoretical realism), I need to consider the traditional replies of the realist to the following questions: (a) 'What is the nature of mathematical reality?', and (b) 'How do we acquire knowledge of mathematical reality?'

These are very important questions indeed, because if one believes that mathematics produces information concerning a certain reality, he then has the obligation to clarify which kind of reality this is (question (a)), and what sort of access he has to it (question (b)). Walking past shrugging his shoulders and leaving these questions unanswered would weaken the realist position. For by enfolding mathematical reality in a noumenal night in which all distinctions are lost, he would expose realism to the accusation of mythologizing.

But there is another reason why questions (a) and (b) are important: addressing them reveals the existence of an interplay between ontology (question (a)) and epistemology (question (b)), an interplay which has a crucial rôle in any philosophy of mathematics, let alone a realist one. For, how can we believe that mathematical entities are objects of a certain kind, if we are in principle unable to give an acceptable account of how we come to know them? And how can we think that we have a plausible account of mathematical knowledge, if we cannot give an account of what this is knowledge of? From these considerations it can be seen that the harmony between mathematical ontology and mathematical epistemology has to be one of the conditions which must be satisfied by any tenable philosophy of mathematics.

In what follows, I shall survey some of what I take to be possible realist answers to questions (a) and (b). But, before I start, a word of warning is in order. One very important thing the reader should keep in mind in going through this chapter is that in writing the sections bearing as titles 'Pythagorean realism', 'Aristotelian realism', and 'Kantian realism' I was not concerned with giving a scholarly presentation of the positions of, respectively, the Pythagoreans, Aristotle and Kant on mathematical realism and on the issues raised by questions (a) and (b). This, given the aim of the book and the enormous amount of relevant literature on these topics, would have been foolish. I simply intended to provide a brief characterization of realist standpoints on questions (a) and (b) which I consider to be relevant to the present-day debate on these issues.

Therefore, the sketches I offer in each section of the positions of the Pythagoreans, Aristotle and Kant are not meant as a contribution to the scholarship concerning the interpretation of the views of these authors, but as an attempt to characterize clusters of ideas on mathematical realism, which, originating from an interpretation of some of these philosophers' positions, have then developed through history into philosophies of mathematics which still have something to offer in the way of answering questions (a) and (b).

2 Pythagorean realism

Pythagoreans are naïve realists about mathematics. They believe that the external world is *mathematical* in the sense that it is, as Galileo puts it, like a book written 'in mathematical language and the characters of this language are triangles, circles, and other geometrical figures without which it is humanly impossible to understand a single word contained in it'. That this is the view characteristic of Pythagoreanism is confirmed by Aristotle who says that[1]

> ...the Pythagoreans, as they are called, devoted themselves to mathematics; they were the first to advance this study, and having been brought up in it they thought its principles were the principles of all things. Since of these principles numbers are by nature the first, and in numbers they seemed to see many resemblances to the things that exist and come into being — more than in fire and earth and water (such and such a modification of numbers being justice, another being soul and reason, another being opportunity — and similarly almost all other things being numerically expressible); since, again, they saw that the attributes and the ratios of the musical scales were expressible in numbers; since, then, all other things seemed in their whole nature to be modelled after numbers, and numbers seemed to be the first things in the whole of nature, they supposed the elements of numbers to be the elements of all things, and the whole heaven to be a musical scale and a number. And all the properties of numbers and scales which they could show to agree with the attributes and parts and the whole arrangement of the heavens, they collected and fitted into their scheme; and if there was a gap anywhere, they readily made additions so as to make their whole theory coherent.

An obvious question that comes to mind at this point is 'Why did the Pythagoreans made such a fuss about numbers vs. other mathematical entities?'. The beginning of an answer to this problem might be that numbers were involved in two of the most important mathematical operations studied at the time, i.e., counting and measuring, and that, perhaps, they had

---
[1] [Aristotle, *Metaphysics*], Book I A, from $985^b 23$ to $986^a 7$.

2. PYTHAGOREAN REALISM

been impressed by a very simple, but striking observation made by several thinkers who came after them — including Frege — according to which, given any concept, it makes sense to ask how many things fall under it.

According to the Pythagoreans, numbers — they only knew the positive integers and ratios of positive integers — were patterns of points. They, in particular, distinguished between triangular numbers, square numbers, oblong numbers, etc. And since they thought that all material objects are sets of points, it is easy to see how they came to believe that all there exists in nature is made out of numbers.

Therefore, for a Pythagorean, mathematical theories are descriptions of entities which exist in the external world or in the objects and structure of the external world (answer to question (a)). Such a view of mathematical reality which, as we have seen in the passage from Aristotle, was in part derived as an explanation of the applicability of mathematics to the empirical sciences, makes it very easy to give an account of how we acquire mathematical knowledge. This is clearly the outcome of our experience of the external world (answer to question (b)).

Having given a summary of the Pythagorean's answers to questions (a) and (b), and having seen that there is harmony between the Pythagorean ideas concerning mathematical ontology and epistemology, the problem that needs to be tackled at this point is whether or not the Pythagorean's view of mathematical reality is tenable. In addressing this question it must be said that, even if we put to one side both the mythological account offered by the 'old Pythagorean' of how the cosmos came about from numbers and the privileged rôle assigned by him to numbers over and above other objects of mathematical investigation, and, following Galileo, keep close to a minimal form of Pythagoreanism according to which mathematical theories, in particular geometry, describe material objects, whereas physics, chemistry, etc. produce explanations and predictions of natural phenomena on the basis of those descriptions,[2] there are very strong objections against the plausibility of the ontological account a Pythagorean philosopher of mathematics can offer of the entities described by mathematical theories.

First, it is possible to object that, in contrast with what asserted by Galileo's Pythagorean manifesto, there are large chunks of the external world which have proved impervious to several attempts to 'read them' using mathematical language. For instance, there are objects in the external world whose shape is too irregular and/or contains too many 'spikes' to become an object of study of Euclidean geometry or even of geometrical theories which make use of the differential calculus. Examples of this kind

---

[2]This seems to me the position defended by Mandelbrot. See [Mandelbrot, 1987], Introduction, pp. 7-8.

are, as Mandelbrot says,[3] the coast-line, the sea, etc.

Of course, to this objection a follower of minimal Pythagoreanism could reply that, in contrast with what it might seem to be the case at first sight, the coast-line, the sea, etc. *are* mathematical objects, they are fractal objects, and we are just beginning to develop (as in the case of Mandelbrot's fractal geometry), the appropriate extension of our mathematical language which, in due course, will enable us to read the still incomprehensible text. But, alas, such an answer would not be compelling, because it boils down to no more than an act of faith in Pythagoreanism.

Secondly, the applicability of numbers to reality in counting objects, measuring their lengths, areas, etc. is not evidence for the fact that numbers are instantiated in concrete objects, or that they are properties of concrete objects, or, as the follower of minimal Pythagoreanism would put it, that numbers describe (properties of) concrete objects, because, for example, what we call 'a book' can be correctly described as one object, or as a set of 552 pages, or as what contains n sentences, m words, k characters, etc. where $1, 552, n, m, k$ are numbers which differ from one another. Therefore, the number we use as a correct description of what occupies a certain portion of space-time changes according to how we 'slice up' such an entity. But, this is not so with regard to the colour, or the shape, or the portion of space-time occupied by the entity in question.

Thirdly, there is evidence that no Euclidean straight line, regular polygon, etc. can be instantiated in concrete objects, or can be used as description of concrete objects. In fact, in the age of the theory of relativity and of quantum mechanics, when it becomes justifiable to represent the universe as having a finite diameter and containing a finite number of elementary particles, and to conceive of space-time as having a curvature different from 0, which concrete object could be considered as a straight line without breadth and infinite in length having 0 curvature, or as an instantiation of a transfinite cardinal number greater than $2^{\aleph_0}$?

The objections above appear to be unanswerable by the Pythagorean philosopher (naïve realist).

3 Aristotelian realism

Aristotle is a realist in the philosophy of mathematics, and when it comes to the problem of explaining the nature of mathematical reality, he says that mathematical entities, e.g., length, surface, etc. are attributes of material things which, very much like attributes such as: motion, white, male, female, do not exist separately from the things they are attributes of:[4]

---

[3]See [Mandelbrot, 1987].
[4][Aristotle, *Metaphysics*], $1078^a 5 - 8$, p. 1704.

3. ARISTOTELIAN REALISM

Many properties attach to things in virtue of their own nature as possessed of some such property; e.g. there are attributes peculiar to the animal *qua* female or *qua* male, yet there is no female nor male separate from animals. And so also there are attributes which belong to things merely as lengths or as planes.

But if mathematical entities are attributes of material things in what sense can Aristotle say that they exist? Are they, as the Pythagoreans thought, immanent in the material bodies? For Aristotle:[5]

> ... the objects of mathematics are not substances in a higher sense than bodies are, ... they are not prior to sensibles in being, but only in formula, and ... they cannot in any way exist separately. But since they could not exist *in* sensibles either, it is plain that they either do not exist at all or exist in a special way and therefore do not exist without qualification. For 'exist' has many senses.

Of course, now the next question is 'In what *relative sense* do mathematical entities exist?'. A reply to this is given by the following quotation:[6]

> Each question will be best investigated in this way — by supposing separate what is not separate, as the arithmetician and the geometer do. For a man *qua* man is one indivisible thing; and the arithmetician supposes one indivisible thing, and then considers whether any attribute belongs to a man *qua* indivisible. But the geometer treats him neither *qua* man nor *qua* indivisible, but as a solid. For evidently the attributes which would have belonged to him even if he had not been indivisible, can belong to him apart from these attributes. Thus, then, geometers speak correctly — they talk about existing things, and their subjects do exist; for being has two forms — it exists not only in fulfillment but also as matter.

According to Aristotle:[7]

> ... matter is just this — the primary substratum of each thing, from which it comes to be, and which persists in the result, not accidentally.

For him, in particular,[8]

> By the matter I mean, for instance, the bronze, by the shape the plan of its form, and by the compound of these (the concrete thing) the statue.[9]

---

[5][Aristotle, *Metaphysics*], $1029^a 19 - 20$, Ch. 3, p. 83.
[6][Aristotle, *Metaphysics*], $1078^a 22 - 30$, Ch. 3, p. 84.
[7][Aristotele, *Physics*], Book I, $192^a 31 - 32$.
[8][Aristotle, *Metaphysics*], $1029^a 3 - 5$, Ch. 3, p. 84.
[9]Concerning the relationship existing between matter and form, Aristotle believes that: [De Ruggiero, 1967], vol. II, Ch. V, §4, pp. 35-36:

> ... the form of an entity is the act by means of which the entity is individuated and determined; its matter is, instead, what can be subjected to such an act, and is in itself mere potentiality, that is, that lack of determination upon which the determining force of the act is exercised. Thus the dyad matter-form intersects the dyad potentiality-act ($\delta\acute{\upsilon}\nu\alpha\mu\iota\varsigma$-'$\varepsilon\nu\acute{\varepsilon}\rho\gamma\varepsilon\iota\alpha$), in which form, in contrast with Platonism, is revealed as an activity which *specifies* matter.

Now, since Aristotle believes that[10]

> By matter I mean that which in itself is neither a particular thing nor of a certain quantity nor assigned to any other of the categories by which being is determined.

it follows that[11]

> ...matter is unknowable in itself.

Nevertheless, we can still say that

> ...some matter is sensible and some intelligible, sensible matter being for instance bronze and wood and all matter that is changeable, and intelligible matter being that which is present in sensible things not *qua* sensible, i.e. in the objects of mathematics.

Therefore, according to the reading of Aristotle I am here presenting, mathematical entities exist as intelligible possibilia which are attributes of the objects of perception and whose knowledge can be attained by abstraction, where abstraction is the operation (the efficient cause) by means of which they are actualized.

Aristotle's view has many merits. On the one hand, by denying that mathematical objects exist in the objects of perception, it avoids the objections we examined in the previous section against the naïve realist. On the other hand, by saying that mathematical objects do not exist independently of the objects of perception, but are attributes, properties of such objects, it opens the way to the construction of a plausible epistemology which would harmonize with its mathematical ontology.

However, independently of the difficulties inherent in the Aristotelian account of mathematical attributes of objects of perception, as opposed to other types of attributes,[12] Aristotle's view of mathematical reality excludes actual infinity from it. And since the concept of actual infinity is at the heart of much of contemporary mathematics, it follows that, if we want to be realists about mathematics as a whole, we should conclude that Aristotle's account of mathematical reality is unsatisfactory.

At this point someone might object that we are proceeding rather too quickly in our criticism of Aristotle's views on mathematical reality. For if, given a mathematical theory T, we were, like Hilbert, prepared to distinguish between statements of T which have a reference (finitary statements) and statements of T which have not (ideal statements) on the basis of the existence of verifiability (or testability) conditions (see §4.3), it seems that

---

[10][Aristotle, *Metaphysics*], $1029^a 20 - 22$, Ch. 3, p. 85.
[11][Aristotle, *Metaphysics*], $1036^a 8 - 11$, Ch. 3, p. 85.
[12]These difficulties are expressed by an obvious reformulation of the second objection against Pythagoreanism (see §3.2).

3. ARISTOTELIAN REALISM

we could hang on to a diminished form of Aristotelianism. This in the sense that we could still consider verifiable (or testable) mathematical statements as giving us information concerning attributes of objects of perception, and, at the same time, view ideal statements (those statements involving actual infinity) as expressions whose meaning is given exclusively in terms of conditions of correct use, and whose importance in mathematics is confined to simplifying and making more efficient and elegant the theory within which they are formulated.

In spite of the persuasiveness of the above argument, this is not compelling. In fact, since the justification for the introduction and use of ideal statements cannot come from their reference to legitimate mathematical entities, this should then at least come from the consideration that, if we were to add the ideal statements of a mathematical theory T to the verifiable/testable/referring statements of T, no contradiction would be generated. But this is precisely the justification — a proof of the absolute consistency of T — which, in the case of many mathematical theories, including set theory, we do not know how to give using methods which are immune to doubts as to their legitimacy (a consequence of Gödel's incompleteness theorems).

An interesting variation on the position examined above is that of D. Gillies. For Gillies:[13]

> Aristotle's support for the potential as opposed to the actual infinite is rather separate from his general analysis of existence in mathematics. One might be able to support the second while denying the first. This is indeed my position.

According to Gillies, sets exist in the material world and since he believes that natural numbers are properties of sets, for him, also natural numbers exist in the material world. Then[14]

> In order to introduce infinite sets into the material world, I will argue that continuous intervals of points are physically real. The points could be of space, or of time, or of space-time. Now when should we regard an entity as physically real, or as a constituent of the physical world? My answer to this question is as follows. Suppose some of the best theories of physics which we have, theories which are very well confirmed by observation and experiment, contain symbols referring to the entity in question, then we should regard that entity as part of the physical world.

For him, both $\aleph_0$ and $2^{\aleph_0}$ occur in the material world, but no cardinal number greater than $2^{\aleph_0}$ does:[15]

> The idea is that the Cantorial theory of the transfinite alephs [greater than $2^{\aleph_0}$] is a metaphysical theory ... That is to say that the theory, although meaningful, is too

---

[13] Private communication.
[14] [Gillies, 2000], p. 49.
[15] [Gillies, 2000], pp. 50-51.

remote from the empirical basis of science to be capable of being either confirmed or disconfirmed by observation and experiment. [And] As for mathematical statements [which] are metaphysical in nature, the concept of truth does not, properly speaking, apply to them.

Gillies's view is an astute attempt to base a robust theory of the actual infinite on Aristotle's view of existence in mathematics *via* a classical indispensability argument. However, it seems to me that Gillies's view can be criticized on the same grounds used to attack the more general position previously examined with regard to the lack of justification for the introduction and use of mathematical statements of metaphysical nature.

But now a die-hard Aristotelian might reply that, taking for granted the compelling nature of the argument above against what we might call the 'Hilbert-Gillies defense', he does not need to be realist about mathematics as a whole. In fact, for him, actual infinity is simply a misguided concept and, consequently, the theories which make use of it are not mathematics, because in them coexist, as Wittgenstein famously said of set theory, 'conceptual confusion and methods of proof.'[16] Of course, such a position would imply a revisionary attitude towards what must be taken as mathematics, but what's wrong with that?

One thing we must notice concerning the position of the die-hard Aristotelian is that he would not stand alone on the issue of actual infinity. He would, in fact, enjoy the company of the small, but loud family of the constructivists (see §§4.7-4.12). This, however, would not be a good company for him to keep, because of the constructivists' marked anti-realist inclinations. Actually, given the profound and pervasive rôle of actual infinity in present day mathematics, it would be correct to say that, by siding with the constructivists on the issue of actual infinity and on the revisionary attitude towards mathematics, the die-hard Aristotelian would have changed sides in the realism/anti-realism dispute in the philosophy of mathematics going from the realist to the anti-realist camp.

4 Kantian realism

For Kant all objects can be distinguished into phenomena and noumena. Phenomena are what is given by sense perception when this is pre-reflectively ordered by the *a priori* pure intuitions of space and time. Phenomena, or what appears to us, are therefore dependent on two things: the entities which have stimulated our senses and our reason's way of structuring perceptual input. Phenomena are, therefore, real and intersubjective.

Noumena, instead, are not objects of sensible intuition. They are objects of intellectual intuition, which is something not given to humankind '... and

---
[16][Wittgenstein, 1983], Part II, §XIV, p. 232^e.

4. KANTIAN REALISM

of which we cannot comprehend even the possibility'.[17]

According to Kant:[18]

> The concept of a *noumenon* — that is, of a thing which is not to be thought as object of the senses but as a thing in itself, solely through a pure understanding — is not in any way contradictory. For we cannot assert of sensibility that it is the sole possible kind of intuition ... The concept of noumenon is ... a merely *limiting concept*, the function of which is to curb the pretensions of sensibility; and it is therefore only of negative employment. At the same time it is no arbitrary invention; it is bound up with the limitation of sensibility, though it cannot affirm anything positive beyond the field of sensibility.

Mathematics is, for Kant, a science of phenomena. But since also physics, like any other science of nature, studies phenomena, it is legitimate to ask (1) how, for Kant, should we distinguish between mathematics and, say, physics, and (2) what sort of phenomenical entities are, for him, the objects of study of mathematics as opposed to the objects of study of physics.

The answer to question (1) is that, according to Kant, a good way of drawing a distinction between mathematics and physics consists in the realization that mathematical judgments are synthetic *a priori*, whereas judgments belonging to physics are synthetic *a posteriori*. But what does this mean?

As is well known, for Kant, judgements are expressed by statements which have a subject-predicate structure, i.e., by statements of the type 'John is bald'. An analytical judgement is, for him, a judgement expressed by a statement in which the predicate does not increase the amount of information contained in the concept of the subject, e.g.: 1) 'The Sun is a body' — in this particular case the concept of body is implicit, *contained* in the definition of Sun.

A synthetic judgement is, on the contrary, a judgement expressed by a statement in which the predicate increases the information contained in the concept of the subject, e.g.: 2) '$\sqrt{2}$ is an irrational number' or 3) 'On the 21st of June 1995, Rome had more than one million inhabitants'. If someone were to give us a definition of Rome as the capital of Italy, etc., we would not be able to derive from this how many inhabitants Rome had on the 21st of June 1995; moreover, saying that $\sqrt{2}$ is the positive real number m such that $m^2 = 2$ does not give any indication as to whether m is rational or irrational. Therefore, asserting that $\sqrt{2}$ is an irrational number and saying that Rome had more than one million inhabitants on the 21st of June 1995 are ways of *extending* the amount of information made available to us simply through an analysis of the definition of $\sqrt{2}$ or of Rome.

---

[17][Kant, 1787], Transcendental Analytic, Ch. III, p. 268.
[18]Ibid., pp. 271-272.

Furthermore, for Kant, a judgement is true (false) *a priori* if it is true (false) independently of experience — the judgement expressed by 1) would be a good example of an *a priori* true judgement, because we would be able to show that it is true simply through an inspection of the definition of Sun; and that expressed by 2) would also be an example of an *a priori* true judgement, because we would be able to determine its truth by producing an abstract proof and not by experience.

Lastly, a judgement is *a posteriori* true (false) if only experience can justify the attribution of truth (falsity) to it: a clear example of *a posteriori* true judgement is represented by that expressed by 3). In fact, only a census taken in Rome on the 21st of June 1995, or a similar empirical verification procedure, is able to confirm or refute such a judgement.

Kant's classification of judgements, and therefore statements, according to the analytic/synthetic and *a priori*/*a posteriory* distinctions allows us to construct a useful classification table in the following way:

	a priori	a posteriori
analytic	1)	
synthetic	2)	3)

Figure 3.1. Judgements Classification Table

In the table above, the box individuated by the pair analytic/*a posteriori* is empty, because, if a judgement is analytically true, then it is true in virtue of the fact that the predicate belongs to the concept of the subject. In other words, if we were to know the concept of the subject, i.e., the list of predicates expressing properties of the subject, then, *independently of experience*, we would realise that the judgement is true, because we would find the predicate (occurring in the judgement) in our list. Therefore, such a judgement could not be *a posteriori*.

If, on the other hand, a judgement is genuinely *a posteriori*, i.e., there is in principle no way of telling whether it is true or false independently of experience, then the judgement could not be analytic, because the list of predicates expressing the properties of the subject would neither contain the predicate occurring in the judgement nor a predicate which is inconsistent with it.

For Kant, mathematical judgements are synthetic *a priori*, because they are true independently of experience and, at the same time, extend the knowledge we can obtain simply by listing the properties of the subject; whereas, judgments belonging to physics, and to the other natural sciences, are synthetic *a posteriori*. Judgments belonging to physics, in fact, are,

4. KANTIAN REALISM

according to Kant, genuinely ampliative, but we need experience (in the form of experiments) to determine whether they are true or false.

If it is true to say that Kant operates a sharp distinction between mathematics and the natural sciences, it is also true to assert that he draws a clear demarcation line between logic and mathematics. With regard to the latter point, it is important to notice that Kant believes that what he calls 'general logic', which is what in his system comes closest to what Frege and the following tradition meant by 'logic', is essentially concerned with the study of the laws of thought; this is a study which is analytic in character, because it:[19]

> ...resolves the whole formal procedure of the understanding and reason into its elements, and exhibits them as principles of all logical criticism of our knowledge.

Furthermore, despite the fact that this part of logic is seen by him as a propaedeutic to all the other sciences, it is not identifiable with them, because although it:[20]

> ...is justified in abstracting — indeed, it is under obligation to do so — from all objects of knowledge and their differences, leaving the understanding nothing to deal with save itself and its form. ... for reason to enter on the sure path of science is, of course, much more difficult, since it has to deal not with itself alone but also with objects. Logic, therefore, as a propaedeutic, forms, as it were, only the vestibule of the sciences; and when we are concerned with specific modes of knowledge, while logic is indeed presupposed in any critical estimate of them, yet for the actual acquiring of them we have to look to the sciences properly and objectively so called.

Although what I have said so far answers question (1), and shows how Kant distinguishes between logic and mathematics, nothing has been said about how Kant answers question (2). Let us now see how he does it.

In Kant's view, our reason is endowed with a spatio-temporal system of representation which produces a pre-reflective ordering of the perceptual input, that is, an ordering of the perceptual input that precedes the reflective activity of reason — this is the activity exercised by reason through the formation of judgements. What this means is that when we have a perception, this is not entirely determined in its properties and structure by the external object(s), but that such perceptory input is spatially and temporally ordered by reason. Moreover, such an ordering is pre-reflective because it is not the consequence of our attempt to interpret and/or understand what we are perceiving; it is simply given to us and determines, to an in principle unspecifiable extent, how things appear to us.

---

[19][Kant, 1787], Transcendental Doctrine of the Elements, Transcendental logic, §III, The Division of General Logic into Analytic and Dialectic, p. 98.
[20][Kant, 1787], Preface to the Second Edition, p. 18.

Now, for Kant, mathematical concepts cannot be *abstracted* from experience, because otherwise the judgements in which they occur could not be known with *a priori* certainty.

Moreover, for any concept and, in particular, for a mathematical concept to be meaningful this has to refer to an object. But if such an object is given to us by our pre-reflective system of representation, how can we possibly have *a priori* knowledge of it? Kant's solution of this problem consists in asserting that the object of mathematical investigation is constructed *a priori* by the mathematician.[21]

> Take ... the concepts of mathematics, considering them first of all in their pure intuitions. Space has three dimensions; between two points there can be only one straight line, etc. Although all these principles, and the representation of the object with which this science occupies itself, are generated in the mind completely *a priori*, they would mean nothing, were we not always able to present their meaning in appearances, that is, in empirical objects. We therefore demand that a bare concept be *made sensible*, that is, that an object corresponding to it be presented in intuition. Otherwise the concept would, as we say, be without *sense*, that is, without meaning. The mathematician meets this demand by the construction of a figure, which, although produced *a priori*, is an appearance present to the senses. In the same science the concept of magnitude seeks its support and sensible meaning in number, and this in turn in the fingers, in the beads of the abacus, or in strokes and points which can be placed before the eyes. The concept itself is always *a priori* in origin, and so likewise are the synthetic principles or formulas derived from such concepts; but their employment and their relation to their professed objects can in the end be sought nowhere but in experience, of whose possibility they contain the formal conditions.

Of course, the question we have to ask next is 'What does Kant mean by "*a priori* construction?"'. Kant addresses this issue in the following quotation:[22]

> *Philosophical* knowledge is the *knowledge gained by reason from concepts*; mathematical knowledge is the knowledge gained by reason from the *construction* of concepts. To *construct* a concept means to exhibit *a priori* the intuition which corresponds to the concept. For the construction of a concept we therefore need a *non-empirical* intuition. The latter must, as intuition, be a *single* object, and yet none the less, as the construction of a concept (a universal representation), it must in its representation express universal validity for all possible intuitions which fall under the same concept. Thus I construct a triangle by representing the object which corresponds to this concept either by imagination alone, in pure intuition, or in accordance therewith also on paper, in empirical intuition — in both cases completely *a priori*, without having borrowed the pattern from any experience. The single figure which we draw is empirical, and yet it serves to express the concept, without impairing its universality. For in this empirical intuition we consider only the act whereby we construct the concept, and abstract from the many determinations (for instance, the magnitude of the sides and of the angles), which are quite indifferent, as not altering the concept 'triangle'.

---

[21] [Kant, 1787], Transcendental Analytic, Analytic of Principles, Ch. III, pp. 259–260.
[22] [Kant, 1787], p. 577.

4. KANTIAN REALISM

Now, as Micheli says, for Kant:[23]

> ...to construct in mathematics does not at all mean to produce an empirical construction; ...to Eberhard who had misunderstood him on this point, Kant replied drawing a clear distinction between 'technical, or empirical, construction' of a geometrical figure, such as for example 'describing a parabola according to the theory', which is a problem that, according to Kant, does not concern the mathematician as such, and the 'pure, or schematic, construction', which consists in determining a method, or a general procedure, which enables us to produce what is thought in a concept, as for example 'obtaining a parabola in a conic section' or constructing it 'analytically' by means of the equation '$ax = y^2$'.

Mathematical constructions are what Kant calls 'schemata', where[24]

> The schema ... is a general rule, a method for determining images according to a pure concept; the schema is not, therefore, the image but the rule for determining images according to concepts. The schema is a product of the intellect: a determination of the way in which we receive objects, which makes it objective. It can be defined as an '*a priori* sensible representation', an *a priori* object, which makes possible to bring the empirical multiplicity under general rules.

The answer to the above question (2) is, therefore, that, since[25]

> Through the determination of pure intuition we can acquire *a priori* knowledge of objects, as in mathematics, but only in regard to their form, as appearances

it follows that[26]

> ...human discursive intellect determines the pure forms of sensible intuition, producing pure mathematical forms, pure formal relationships, forms of possible objects, under which all empirical data can in principle be brought.

In other words, mathematical objects are, for Kant, schemata (not images) of possible phenomena, where 'possible' means compatible with the conditions for experience (and not simply non-contradictory).[27]

---

[23][Micheli, 1998], Ch. 4, pp. 30-31.
[24][Micheli, 1998], ibid., §d p. 43.
[25][Kant, 1787], Transcendental Analytic, Book I, Ch. II, Transcendental Deduction (B), §22, p. 162.
[26][Micheli, 1998], ibid., §c p. 41.
[27][Kant, 1787], ibid., Book II, Ch. I, p. 182:

> If five points be set alongside one another, thus, ..., I have an image of the number five. But if, on the other hand, I think only a number in general, whether it be five or a hundred, this thought is rather the representation of a method whereby a multiplicity, for instance a thousand, may be represented in an image in conformity with a certain concept, than the image itself. For with such a number as a thousand the image can hardly be surveyed and compared with the concept. This representation of a universal procedure of imagination in providing an image for a concept, I entitle the schema of this concept.
> Indeed it is schemata, not images of objects, which underlie our pure sensible concepts. No image could ever be adequate to the concept of a triangle in general.

Before proceeding to a critical evaluation of Kant's position on mathematics, it is very important to bring to the reader's attention some features of the view just expounded.

First, although, as we have seen, for Kant, mathematical concepts are constructed, i.e., we are able to exhibit *a priori* the intuition which corresponds to them, and mathematical objects are schemata, i.e., general rules for determining images according to a pure concept, Kant is neither a conventionalist nor a constructivist about mathematics.

He is not a conventionalist, because, for him, mathematical concepts are not arbitrary, or established simply by convention, because they refer to objects which are the schemata of possible phenomena, i.e., of phenomena which are compatible with the conditions of experience. And he is not a constructivist like Brouwer either, because, as already emphasized, Kant is a realist about mathematics and his talk of 'exhibiting *a priori* the intuition which corresponds to mathematical concepts' and of 'general rules for determining images according to a pure concept' has to do with the process of concept formation in mathematics and with the nature of mathematical objects and not with mathematical proof.

Moreover, Kant's account of the process of concept formation in mathematics and of the nature of mathematical objects is not matched, like in Brouwer, by a sharp criticism of the use of non-constructive proof procedures in mathematics and by the consequent conviction that mathematics needs reforming in a constructivist direction.

Of course, to this a Brouwerian intuitionist might reply that (a) Brouwer's Intuitionism is a faithful realization of Kant's views on mathematics, because a constructive proof in Brouwer's sense does precisely the job of 'exhibiting *a priori* the intuition which corresponds to mathematical concepts'[28]; and that (b) Brouwer's well-known idea concerning the irrelevance

---

> It would never attain that universality of the concept which renders it valid of all triangles, whether right-angled, obtuse-angled, or acute-angled; it would always be limited to a part only of this sphere. The schema of the triangle can exist nowhere but in thought. It is a rule of synthesis of the imagination, in respect to pure figures in space.

and ([Kant, 1787], ibid., pp. 183-184)

> The pure image of all magnitudes (*quantorum*) for outer sense is space; that of all objects of the senses in general is time. But the pure *schema* of magnitude (*quantitatis*), as a concept of the understanding, is *number*, a representation which comprises the successive addition of homogeneous units. Number is therefore simply the unity of the synthesis of the manifold of a homogeneous intuition in general, a unity due to my generating time itself in the apprehension of the intuition.

[28]This is particularly clear in the case represented by what it means, for Brouwer, to give a constructive proof of $\exists x \phi(x)$. According to Brouwer, a constructive proof of $\exists x \phi(x)$ is a finite *a priori* procedure which individuates (exhibits, gives a representation

4. KANTIAN REALISM

of language to mathematical activity,[29] far from being a loud and clear expression of his idiosyncratic tendency towards psychologism, is precisely the realization of the kantian idea that, if we cannot exhibit *a priori* the intuition which corresponds to a mathematical concept, then we have no mathematical concept in the first place.

Now, it is undeniable that the Brouwerian intuitionist's reply has some force. It is, in fact, the case that a constructive proof in Brouwer's sense has this *positive* feature of, for instance, producing an *a priori* procedure which is able to exhibit the object we are claiming the existence of as falling under a certain concept. What, however, in the absence of the slightest hint made by Kant in this direction, does not appear to me to be correct in the reply of the Brouwerian intuitionist is the idea that, for Kant, mathematical proofs are/should be constructive in Brouwer's sense.

Secondly, the idea of mathematical entities as schemata, and not as images, on the one hand, saves the objectivity of mathematics and, on the other hand, highlights the abstract, relational nature of the objects of mathematical investigation.

It saves the objectivity of mathematics, because if mathematical entities were images they would be private objects in Frege's sense. In fact, if mathematical entities were images, how could I possibly know whether the image of a triangle I have right now is the same as yours or the same as I had in the past when I saw/imagined a triangle? But since, for Kant, mathematical entities are rules for the determination of images according to concepts, there is no difficulty in giving an objective specification of such rules. Furthermore, the nature of mathematical entities is abstract and relational, because they generate pure forms of possible objects of empirical intuition.

Thirdly, since Kant's position on mathematical reality does not place the objects of investigation of mathematics in the external world, the problem of explaining how access to mathematical objects is obtained has a trivial solution: they are a product of the intellect.

Fourthly, this view disposes of the problems that Pythagoreans and Aristotelians suffer from. Mathematical objects are neither in the objects of perception nor are they properties (attributes) of such objects. Moreover, actual infinity is easily accounted for by Kant, because it is guaranteed by the properties of the *a priori* intuition of space:[30]

---

of) an element d of the domain of quantification D and shows that d has the property ϕ.

[29] For Brouwer, if we have a mathematical statement for which we do not yet have a constructive proof or a refutation (a conjecture), this must be considered as mathematically meaningless.

[30] [Kant, 1787], Transcendental Æsthetic, §2, pp. 69-70.

94   TRADITIONAL REALIST POSITIONS

> Space is represented as an infinite *given* magnitude. Now every concept must be thought as a representation which is contained in an infinite number of different possible representations (as their common character), and which therefore contains these *under* itself; but no concept, as such, can be thought as containing an infinite number of representations *within* itself. It is in this latter way, however, that space is thought; for all the parts of space coexist *ad infinitum*. Consequently, the original representation of space is an *a priori* intuition, not a concept.

However, in spite of the plausibility of the general lines of Kant's position on mathematical knowledge and reality, which display a beautiful harmony between ontology and epistemology, and its advantages over Pythagorean and Aristotelian realism, there are several problems that arise. I shall highlight some of these in what follows.

According to Kant, mathematical judgements (i) are true independently of experience (*a priori*), and (ii) are ampliative (synthetic); moreover, (iii) negating true mathematical judgements causes contradictions to appear (they are necessary),[31] and (iv) no exceptions to a true mathematical judgement are allowed as possible (they are universal).

What, for Kant, is, in particular, at the root of the *a priori*, necessary, and universal nature of geometrical statements is the[32]

---

[31] Condition (iii) does not imply that mathematical statements are analytic in Kant's sense, because the appearing of a contradiction from negating, for example, the statement $2 + 2 = 4$ does not show that the concept of 4 is contained in the concept of $2 + 2$. To see this consider the assertion $2 + 2 = 3$ (the other possible cases can be dealt with in a similar way). And imagine also that, to make things easier, we work within Peano arithmetic — Peano axioms are given in footnote 39. Then, if n is a natural number and n' is the immediate successor of n, we have that:

$$(+) \quad \text{if } m, n \in \mathbb{N} \text{ then } m + n = \begin{cases} m, & \text{if } n = 0 \\ (m + (n-1))', & \text{if } n \neq 0. \end{cases}$$

$$
\begin{align}
2 + 2 &= 3 \quad \text{assumption} \tag{3.1}\\
2 + 2 &= 2' \quad (+) \tag{3.2}\\
2'' &= 2' \quad (+) \tag{3.3}\\
2' &= 2 \quad \text{Peano axiom (3) applied to (3.3)} \tag{3.4}\\
2' &= 1' \quad (+) \tag{3.5}\\
2 &= 1 \quad \text{Peano axiom (3) applied to (3.5)} \tag{3.6}\\
1' &= 0' \quad (+) \tag{3.7}\\
1 &= 0 \quad \text{Peano axiom (3) applied to (3.7)} \tag{3.8}\\
0' &= 0 \quad (+). \tag{3.9}
\end{align}
$$

Since 0 is a natural number (Peano axiom (1)), (3.9) contradicts Peano axiom (4) □ In the reasoning leading to the contradiction the concept of 4 has not been mentioned even once.

[32] Ibid., pp. 68-69.

4. KANTIAN REALISM

... *a priori* necessity of space.³³ Were this representation of space a concept acquired *a posteriori*, and derived from outer experience in general, the first principles of mathematical determination would be nothing but perceptions. They would therefore all share in the contingent character of perception; that there should be only one straight line between two points would not be necessary, but only what experience always teaches. What is derived from experience has only comparative universality, namely, that which is obtained through induction. We should therefore only be able to say that, so far as hitherto observed, no space has been found which has more than three dimensions.

But, of course, with the advent of non-Euclidean geometries and of n-dimensional, for $n > 3$, metric spaces, the *a priori* necessity of 3-dimensional Euclidean space — which is Kant's idea of space in the *Critique of Pure Reason* — has to go, and with it must also go 'The apodeictic certainty of all geometrical propositions, and the possibility of their *a priori* construction'.

Frege defends Kant's position asserting that:³⁴

> Empirical propositions hold good of what is physically or psychologically actual, the truths of [3-dimensional Euclidean] geometry govern all that is spatially intuitable, whether actual or product of our fancy. The wildest visions of delirium, the boldest inventions of legend and poetry, where animals speak and stars stand still, where men are turned to stone and trees turn into men where the drowning haul themselves up out of swamps by their own topknots — all these remain, so long as they remain intuitable, still subject to the axioms of [3-dimensional Euclidean] geometry. Conceptual thought alone can after a fashion shake off this yoke, when it assumes, say, a space of four dimensions or positive curvature. To study such conceptions is not useless by any means; but it is to leave the ground of intuition entirely behind.

It seems to me that what I am going to call 'Frege's defense' boils down to saying that Euclidean geometry is necessary for pre-reflective spatial representation to take place, and that it is acceptable, for Kant, to hold that 'Conceptual thought alone can after a fashion shake off this yoke' of intuition in the development of non-Euclidean geometries.

Frege's defense, although rather astute, does not quite work. In fact, if, on the one hand, Frege is right in claiming that the type of necessity associated by Kant to Euclidean geometry is that of a precondition of experience — if you imagine something spatially then you must imagine it in a Euclidean 3-d space — it is, on the other hand, wide of the mark to think, as Frege does, that it makes sense, for Kant, to talk about mathematical activity as conceptual thought that can leave the ground of intuition entirely behind.

---

³³[Kant, 1787], ibid., p. 68:

> Space is a necessary *a priori* representation, which underlies all outer intuitions. We can never represent to ourselves the absence of space, though we can quite well think it as empty of objects. It must therefore be regarded as the condition of the possibility of appearances, and not as a determination dependent upon them. It is an *a priori* representation, which necessarily underlies outer appearances.

³⁴[Frege, 1884], §14, p. 20.

The main reason in support of the latter point should be clear by now. As we have already seen, according to Kant, mathematical concepts are constructions and 'To *construct* a concept means to exhibit *a priori* the intuition which corresponds to the concept'. Therefore, for Kant, it does not make any sense to talk about a mathematical thought that leaves intuition entirely behind.

As a matter of fact, it is also rather dubious what Kant, and Frege in his defense of Kant, appear to be taking for granted, i.e., that if we imagine something spatially, we must imagine it in Euclidean space. Indeed, as Helmholtz, and others convincingly argue (see §2.7), it appears to be the case that 3-dimensional Euclidean geometry is not a necessary condition for pre-reflective spatial representation to take place.

Lastly, Frege, criticizing Kant in *The Foundations of Arithmetic*, asserts that if mathematical statements are synthetic *a priori*[35]

> ...there is no alternative but to invoke a pure intuition as the ultimate ground of our knowledge of such judgments, hard though it is to say of this whether it is spatial or temporal, or whatever else it may be.

But, says Frege, since intuition in the *Critique of Pure Reason* has to do with sensibility, it follows that 'an intuition in this sense cannot serve as the ground of our knowledge of the laws of arithmetic',[36] because 'I cannot even allow an intuition of 100,000, far less of number in general, not to mention magnitude in general'.[37]

In reading this passage, one has the impression that Frege thinks that for Kant an intuition of 100,000 is some kind of image. This he believes to be absurd, because we cannot even visualize such a multeplicity. But, as we have already seen, Kant does not believe that the intuition corresponding to the concept of a number is an image. The crucial passages relating to this question are the quotations from [Kant, 1787] given in footnote 27. There Kant distinguishes very clearly between having an image of a number and thinking a number in general. Thinking a number in general, whatever its magnitude may be, means, for Kant, having an intuition/representation of the method whereby the multeplicity which we are thinking may be represented in an image, and not the image itself. For example, thinking 100,000 is having a representation of a method — adding one entity (a dot or a stroke or ...) to a set of homogeneous entities (dots or strokes or ...) — which will represent the concept in an image (sets of dots, strokes, etc.) corresponding to it. Furthermore, Kant in these passages attracts the at-

---

[35] [Frege, 1884], §12 pp. 17-18.
[36] [Frege, 1884], §12 pp. 19.
[37] [Frege, 1884], ibid.

5. *PLATONISM*　　　　　　　　　　　　　　　　　　　　　　　　　　　　97

tention of the reader precisely to the problems concerning the surveyability of images. The same considerations, he shows, apply to geometry.

5　Platonism

'Platonism' is a word which refers to a family of doctrines whose common factor is the belief that mathematical reality differs from that of the objects belonging to the external world, and from that of mental states. For the Platonist, mathematical reality is made of abstract entities. But, perhaps, to avoid possible sources of misunderstanding I ought to be more precise about these matters.

If by 'object' we mean what exists independently of whether we are thinking about it, the Platonist draws a distinction between concrete and abstract objects. A concrete object is what occupies a particular portion of space-time; whereas an abstract object does not have such a property. An example of concrete object is given by the typographical characters contained in a copy of this book, whereas the number describing how many of these there are is an example of abstract object.

One of the characteristic features of concrete objects is that they are bound by the portion of space-time they occupy in the sense that a concrete object cannot be in two different places at the same time. Of course, we could have several copies of this book, but these would be concrete objects which would differ from one another.

On the other hand, the same number can correctly describe the quantity of characters contained in a wide variety of items occupying different portions of space-time from one another, as in the case in which there are multiple copies of the same book.

Now that we are in possession of an explicit way of distinguishing between concrete and abstract objects, it is important to point out that the main difference existing between the various Platonist views of mathematics lies in the contrasting characterizations offered of mathematical reality.

In what follows, I am going to study Platonism with the aim of assessing its answers to questions (a) and (b) formulated in the introduction by means of an analysis of four Platonist views concerning what the abstract entities which represent the object of study of mathematics are. But before I do so, let me briefly examine one of the motivations which have made of this position the most popular among the realist views of mathematics.

The Platonist, in contrast with the naïve realist and the Aristotelian realist, has no difficulty in explaining why mathematical entities are not realized in concrete objects. Indeed, if mathematics is a science of 'relations of ideas' (Hume) or of 'relations between thoughts' (Frege), and not of matters of fact, there is no reason to believe that abstract entities such as ideas or thoughts be realized in concrete objects, nor to expect of coming to know about or refer to such entities through the establishing of causal chains. Abstract objects are not subject to the principle of action and reaction.

Therefore, Platonism, can very easily handle the problem posed by an attempt to justify a realist account of, among other things, actual infinity and non-Euclidean geometries.

6 Platonism about objects

The oldest form of Platonism is the so-called 'Platonism about objects', and a traditional form of Platonism about objects is the view according to which numbers are abstract objects, and number theory is the study of their properties. The set-theoretical variation on this theme simply adds to what has just been said the idea that numbers are sets.

Before proceeding further in the discussion of this type of Platonism, we must consider that, for a Platonist about objects, abstract mathematical objects, like concrete objects, have properties which determine the way in which they relate to each other, and not *vice versa*. For example, it is the property of 3 of being the cardinal number of the union of a set of two elements A with a set of one element B, where $A \cap B = \emptyset$, what explains why 3 is the immediate successor of 2, and not the other way round. Therefore, for a Platonist about objects, it makes sense to talk about the identity conditions of, say, the number 3 independently of the number 3 being related to the other natural numbers in such-and-such a way. This, apart from being a defining property of Platonism about objects, is also at the root of the belief that the Platonist about objects has in the uniquenesss of, for instance, the natural numbers.

The intuition behind the most important objection against Platonism about objects goes back to Russell,[38] but I will here examine a set-theoretical version of it formulated by Paul Benacerraf.

We know that arithmetic can be axiomatized by the familiar 5 axioms of Peano[39] using the 3 primitive notions of 0, number, and successor; and

---

[38][Russell, 1993], Chapter I.
[39][Russell, 1993], Ch. I, pp. 5-6:

(**1**) 0 is a number.

(**2**) The successor of any number is a number.

(**3**) No two numbers have the same successor.

(**4**) 0 is not the successor of any number.

(**5**) Any property which belongs to 0, and also to the successor of every number which has the property, belongs to all numbers.

6. PLATONISM ABOUT OBJECTS

that arithmetic can be expressed within set theory in (at least) two different ways, call them the 'Zermelo way' (Z) and the 'von Neumann way' (vN). (See fig. 3.2)

$$
\begin{array}{ccccc}
 & \emptyset, & \{\emptyset\}, & \{\{\emptyset\}\}, & \cdots \\
(Z_n) & \uparrow & \uparrow & \uparrow & \uparrow \\
 & 0, & 1, & 2, & \cdots \\
(vN_m) & \downarrow & \downarrow & \downarrow & \downarrow \\
 & \emptyset, & \{\emptyset\}, & \{\emptyset, \{\emptyset\}\}, & \cdots
\end{array}
$$

Figure 3.2.

Since the sequences of Z-sets and vN-sets[40] are equally adequate to satisfy the set-theoretical version of Peano's axioms,[41]

> ...the fact that [(Z_n) and (vN_m)] disagree on which particular sets the numbers [k, for $k > 1$] are [because $Z_k \neq vN_k$, for $k > 1$] is fatal to the view that each number is some particular set.

But if, as a consequence of Benacerraf's objection, it makes no longer sense to say that each number is some particular set then the set-theoretical version of Platonism about objects is under threat. For in what sense could we say that arithmetic is the study of the properties of certain sets if we are not able to say which sets these are, that is, whether they are Z-sets or vN-sets?

It is important to notice that the phenomenon expressed by Benacerraf's objection is not something confined to set-theoretical representations of number theory. In fact, as is well known, given a particular consistent formal mathematical system, the objects to which this system refers can be determined only up to isomorphism. In other words, given a particular consistent axiomatic system $\mathcal{G}$, there are at least two isomorphic models $\mathcal{M}_1$ and $\mathcal{M}_2$ of $\mathcal{G}$ such that the objects belonging to the domains of $\mathcal{M}_1$ and $\mathcal{M}_2$ differ from one another. What this implies is that it is impossible, from simply considering $\mathcal{G}$ and its models, to determine what sort of objects $\mathcal{G}$ is about.

---

[40] If (Z_n) is the sequence of Z-sets, we have that:
$$Z_0 = \emptyset \text{ and } Z_{n+1} = \{Z_n\};$$
and if (vN_m) is the sequence of vN-sets then:
$$vN_0 = \emptyset \text{ and } vN_{n+1} = vN_n \bigcup \{vN_n\}.$$

[41] [Benacerraf, 1985], p. 279.

100   TRADITIONAL REALIST POSITIONS

Moreover, since for the Platonist mathematical objects are abstract, i.e., they do not occupy a portion of space-time, there is no possibility of empirically individuating them, say, in the way we would be able to individuate or empirically check the existence of a table. These considerations imply that there is no way at all of identifying what sort of objects $\mathcal{G}$ describes the properties of.

To this objection the Platonist may reply that he has some kind of access to the abstract objects of which $\mathcal{G}$ is a satisfactory description, i.e., the objects belonging to the domain of the intended interpretation of the language of $\mathcal{G}$, and that the only fact shown by the existence of a number n, for $n > 1$, of isomorphic models of $\mathcal{G}$ is that we can *translate* $\mathcal{G}$ into different languages preserving the truth of all the statements that are provable in $\mathcal{G}$.

However, this point may be challenged by the anti-realist, who can say that the argument supplied is unacceptable, because, unless the Platonist produces a clarification of how he can access the realm of abstract objects, his talk about an intended interpretation of $\mathcal{G}$ is bound to remain unconvincing. In other words, the anti-realist is here saying that there is a gap in the Platonist's argument caused by a disharmony existing between the Platonist's ontology of abstract objects and his lacking epistemology.

At this juncture in the debate it seems that there are two possible strategies that the Platonist can adopt to respond to the anti-realist. The first aims at preserving Platonism about objects by giving an account of the accessibility conditions to the realm of abstract objects that mathematics is about. The second consists, instead, in giving up Platonism about objects in favour of a different form of Platonism which does not fall victim to the same objections as Platonism about objects.

7 Fregean platonism

One of the best known Platonist attempts to give an account of the accessibility conditions to the realm of the natural numbers is Frege's.

For Frege, natural numbers are objects, but they are neither ideas (because they are objective) nor, in contrast with the objects of investigation of geometry, are they object of intuition, because[42]

> In arithmetic we are not concerned with objects which we come to know as something alien from without through the medium of the senses, but with objects given directly to our reason and, as its nearest kin, utterly transparent to it.

But now the most pressing question becomes:[43]

> How, then, are numbers to be given to us, if we cannot have any ideas or intuitions of them?

---

[42][Frege, 1884], §105, p. 115.
[43][Frege, 1884], §62, p. 73.

7. FREGEAN PLATONISM

In his attempt to answer the question above, Frege, first of all, argues that the view on the meaning of words according to which if we can form no idea/picture of the content of a word such a word has no meaning depends on believing that the meaning of a word can be determined considering the word in isolation, i.e., independently of the propositional context in which it occurs. But, for Frege, such a belief must be considered to be manifestly false, so much so that he puts its negation, the so-called 'context principle', as one of the three fundamental principles on which his whole enquiry in *The Foundations of Arithmetic* rests:[44]

> never to ask for the meaning of a word in isolation, but only in the context of a proposition

And, indeed, it is the context principle that allows him to solve what we might call the 'accessibility problem'. Frege does this in the following way. He argues that:[45]

> Only in a proposition have the words really a meaning. It may be that mental pictures float before us all the while, but these need not correspond to the logical elements in the judgment. It is enough if the proposition taken as a whole has a sense; it is this that confers on its parts also their content.
>
> This observation is destined, I believe, to throw light on quite a number of difficult concepts, among them that of the infinitesimal, and its scope is not restricted to mathematics either.
>
> The self-subsistence which I am claiming for number is not to be taken to mean that a number word signifies something when removed from the context of a proposition, but only to preclude the use of such words as predicates or attributes, which appreciably alters their meaning.

Now,[46]

> Since it is only in the context of a proposition that words have any meaning, our problem becomes this: To define the sense of ... [those propositions] which express our recognition of a number as the same again. If we are to use the symbol a to signify an object, we must have a criterion for deciding in all cases whether b is the same as a, even if it is not always in our power to apply this criterion.

Therefore, for Frege, natural numbers are given to us in language. And we can determine their identity conditions through defining the sense of propositions 'which express our recognition of a number as the same again'.[47] These are propositions such as 'the number which belongs to the concept F is the same as that which belongs to the concept G'.[48]

---

[44][Frege, 1884], Introduction, p. X.
[45][Frege, 1884], §60, pp. 71-72.
[46][Frege, 1884], §62, p. 73.
[47][Frege, 1884], §62, p. 73.
[48][Frege, 1884], §62, p. 73.

Following this line of thought, Frege formulated in *The Foundations of Arithmetic* the proposition which spells out the identity conditions for numbers. According to such a proposition — called by C. Wright '$N^=$' and by G. Boolos 'Hume's Principle'—

> The number of Fs = the number of Gs if and only if there is a one-one map of the Fs onto the Gs.[49]

Having done so, Frege immediately realized that $N^=$ was unequal to the task it was supposed to achieve, because, without a previous knowledge of what a number is, $N^=$ could not determine the truth-value of propositions like

> The number of Fs = Julius Caesar.

As a consequence of this, Frege decided to provide an explicit definition of number:[50]

> the Number which belongs to the concept F is the extension of the concept 'equal to the concept F'

which allowed him to solve the 'Julius Caesar' problem, and from which he then obtained $N^=$. In other words, for Frege, we are justified in believing in numbers as objects, even though these are neither ideas nor objects of intuition, because, being in possession of non-vague identity conditions for numbers, we can say true (or false) things about them.

The move made by Frege to solve the accessibility problem had momentous consequences. As Michael Dummett has correctly pointed out, Frege's proposed solution of the accessibility problem marks the date of birth of the linguistic turn and with it of analytical philosophy. In fact, since it is only through language that we gain access to what Frege in *Der Gedanke* called the 'realm of thought',[51] it follows that for Frege and, consequently, for the

---

[49]$N^=$ is commonly known as an *abstraction principle*, because it legitimizes the use of an abstract term forming operator: The number of Fs.

[50][Frege, 1884], §68, pp. 79-80.

[51]Logic, for Frege, is the discovery/study of the laws of truth, which, in turn, are also the laws of thought, and arithmetic is seen by him as part of logic [Frege 1977b,], pp. 1-2:

> From the laws of truth there follow prescriptions about asserting, thinking, judging, inferring. And we well speak of laws of thought in this way too. But there is at once a danger here of confusing different things. People may very well interpret the expression 'law of thought' by analogy with 'law of nature' and then have in mind general features of thinking as a mental occurrence ... In order to avoid any misunderstanding and prevent the blurring of the boundary between psychology and logic, I assign to logic the task of discovering the laws of truth, not the laws of

7. FREGEAN PLATONISM 103

analytical philosopher the study of language becomes prior, in the order of the explanation, to the study of thought.

However, with the failure of the logicist programme fails also the idea that in arithmetic, as in logic, we are concerned '...with objects given directly to our reason ...[which] as its nearest kin, [are] utterly transparent to it' and, consequently, our Platonist about objects can no longer rely on the language/concept road to number[52] and is back to square one having to face up to the accessibility problem.

Before moving on to the study of other types of Platonism about objects which offer interesting solutions to the accessibility problem, it is important to point out that in recent years Frege's Logicism has been somewhat revived by various authors, in particular, by C. Wright.

According to Wright, if the Frege of *The Foundations of Arithmetic* had resisted the temptation to give an explicit definition of (cardinal) number in terms of classes holding on to $N^=$ considering it as a stipulation, he would have avoided the contradictions realizing his logicist programme. But let us have Wright himself stating his position:[53]

> ...the neo-Fregean thesis about arithmetic is that a knowledge of its fundamental laws (essentially, the Dedekind-Peano axioms) — and hence of the existence of a range of objects which satisfy them — may be based a priori on the explanatory principle, $N^=$. More specifically, the thesis involves four ingredient claims:
>
> (i) that the vocabulary of higher-order logic plus the cardinality operator, 'Nx:...x ...', provides a sufficient definitional basis for a statement of the basic laws of arithmetic;
>
> (ii) that when they are so stated, $N^=$ provides for a derivation of those laws within higher-order logic;
>
> (iii) that someone who understood a higher-order language to which the cardinality operator was to be added would learn, on being told that $N^=$ is analytic of that operator, all that it is necessary to know in order to construe any of the new statements that would be formulable.
>
> (iv) Finally, and crucially, that $N^=$ may be laid down *without significant epistemological obligation*: that it may simply be stipulated as an explanation of the meaning of statements of numerical identity, and that — beyond the issue of the satisfaction of the truth-conditions it thereby lays down for such statements — no competent demand arises for an independent assurance that there *are* objects whose conditions of identity are as it stipulates.

Concerning the correctness of Wright's claim there has been much controversy.

---

taking things to be true or of thinking.

[52] For Frege '...the content of a statement of number is an assertion about a concept' ([Frege, 1884], §46, p. 59).
[53] [Wright, 1998], §I, p. 389.

First, M. Dummett objects to **(iv)** that[54]

> In *Grundgesetze*, value-ranges are introduced in a manner precisely analogous to that in which Wright argued, in his book,[55] that Frege ought to have introduced cardinal numbers (save that it involved a more explicit means of resolving the Julius Caesar problem): and yet it was so far from being justified as to lead to actual contradiction. It therefore *could* not be maintained that this procedure is, in and of itself, legitimate. Possibly some restriction, distinguishing the case of cardinal numbers from that of value-ranges, could be framed. But Wright had put forward no such restriction; and hence his thesis, as it stood, could not be sound.

Wright, in his reply to Dummett's objection, acknowledges that the restriction he, in fact, suggests to impose upon abstraction principles is not sufficient to ensure consistency.[56] But, he says, there is no need for such a water-tight restriction, because there is no reason why a 'rotten apple effect' starting from Axiom V of *Grundgesetze* should cast doubts concerning the consistency of the whole basketful of abstraction principles and, in particular, concerning the consistency of $\mathbf{N}^=$.

Now, it seems to me that Wright's reply to Dummett's objection is not satisfactory, because the act of postulating an abstraction principle, above all in the presence of abstraction principles analogous to it which are provably inconsistent, seems to have, as Russell famously said, only the advantages of theft over honest toil.

Secondly, the incompleteness of first-order arithmetic shows that both Frege's and Wright's arguments in favour of Logicism cannot work. Frege's attempt to justify Logicism cannot work, because its main argumentative strength resides on the exploitation of formal systems. In fact, if there were no formal system of arithmetic, it would be impossible to show that arithmetical statements are analytic in Frege's sense. For, in the absence of a formal system of arithmetic, how could one show that there exists a general procedure for transforming the proofs of arithmetical theorems into proofs of the same theorems 'from logic alone'?

On the other hand, if arithmetic can be axiomatized by a consistent and complete set of axioms, it is sufficient to prove such axioms from logic alone to show that arithmetical statements are analytic in Frege's sense. But, as is well known, Gödel's first incompleteness theorem shows that any consistent formal system of first-order arithmetic whose set of axioms is recursive is incomplete. From this follows that Frege cannot prove that Logicism is right. The same type of argument is applicable to Wright's neo-Fregean position

---

[54][Dummett, 1998], §3, p. 375.
[55][Wright, 1983].
[56]Frege's modification of Axiom V of *Grundgesetze* is inconsistent, but goes through Wright's restriction.

which makes an explicit reference to the Peano axioms as the fundamental laws of arithmetic.

Thirdly, in contrast with what Wright says in (iv), the existence of different models of Peano Arithmetic lays down a significant epistemological obligation towards clarifying what we are talking about when we speak of natural numbers.

Indeed, the well known fact that the Peano axioms are true of any progression generates an inscrutability of reference phenomenon which strongly motivates a competent demand for both an independent assurance that there are objects — the natural numbers — characterized by the Peano axioms whose identity conditions are as $\mathbf{N}^=$ stipulates, and for a way of making sure that, when we use singular terms within arithmetic, we are referring to such objects, and not to others.

Lastly, as Parsons has noticed,[57]

> ...given the power of second-order logic, it would be difficult to maintain that this derivation [of the axioms of **PA** from Hume's Principle in second-order logic] obtains the elementary Peano Axioms from something more evident.

In actual fact, an appeal to second-order logic appears to threaten at the root not only the idea that a reduction of arithmetic to logic should increase the evidence we have for believing in the truth of **PA**, but the Fregean project altogether. For, if, as Quine said, 'second-order logic is set theory in sheep's clothing', a derivation of Peano Axioms from Hume's Principle in second-order logic would not show that arithmetical statements are analytic.[58]

8 Mathematics as an eidetic science

A now classical Platonist attempt to answer in a most general setting the worries expressed by what I might call the 'accessibility objection' is that offered by Husserl's philosophy of mathematics. In what follows I shall briefly analyze two of its basic concepts, the concepts of essence, and of intuition of an essence, which have a direct bearing on the question at issue.

According to Husserl the word 'essence' refers to:[59]

> ...what is to be found in the very own being of an individuum as the What of an individuum.

Essences are by him considered not to be matters of fact in Hume's sense. They are not entities given in spacetime, but are rather abstract objects.

---

[57][Parsons, 2000], §2, p. 305.
[58]See on this [Martino, 2004].
[59][Husserl, 1998], First Book, Part one, Chapter one, §3, p. 8.

Husserl's reasons for calling essences, and in particular numbers, 'objects' are that (i) essences, like concrete objects, are inexhaustible, because:[60]

> ...they not only have a richness of features that we know, but also are conceived of as having much more to them than what we know, something that we might want to find out

and (ii) essences '...can be made the subject of true and false predications ...'.[61] With regard to the special case represented by the nature of the entities (essences) studied by the mathematician, Husserl says that[62]

> ...the *geometer* ...explores not actualities but "ideal possibilities", not predicatively formed actuality-complexes but predicatively formed eidetic affair-complexes

But what do we have to understand by it? Here Husserl's 'actual/ideally possible' distinction appears to match the distinction drawn by Frege in 'Thoughts' between actual and real. In fact, for Frege[63]

> A thought, admittedly, is not the sort of thing to which it is usual to apply the term 'actual'. The world of actuality is a world in which this acts on that and changes it and again undergoes reactions itself and is changed by them. All this is a process in time. We will hardly admit what is timeless and unchangeable to be actual. Now is a thought changeable or is it timeless? The thought we express by the Pythagorean theorem is surely timeless, eternal, unvarying.

As we have just seen, thoughts, for Frege, are not actual; and, therefore, cannot be matters of fact either. However, according to Frege, thoughts are real:[64]

> ...thoughts are neither things in the external world nor ideas.
>
> A third realm must be recognized. Anything belonging to this realm has it in common with ideas that it cannot be perceived by the senses, but has it in common with things that it does not need an owner so as to belong to the contents of his consciousness. Thus for example the thought we have expressed in the Pythagorean theorem is timelessly true, true independently of whether anyone takes it to be true. It needs no owner. It is not true only from the time when it is discovered; just as a planet, even before anyone saw it, was in interaction with other planets.[65]

The idea of an Husserlian realm of essence, which is very much like Frege's Third Realm, is strengthened by the consideration that: (1) in talking about 'ideal possibilities' Husserl does not intend to deny the reality of mathematical essences, but to characterize such a reality by means of a predicate —

---

[60][Føllesdal 1999,], p. 390.
[61][Husserl, 1998], ibid., p. 10.
[62][Husserl, 1998], ibid., §7, p. 16.
[63][Frege 1977b,], p. 27.
[64][Frege 1977b,], pp. 17-18.
[65]A person sees a thing, has an idea, grasps or thinks a thought he does not create it but only comes to stand in a certain relation to what already existed — a different relation from seeing a thing or having an idea.

8. MATHEMATICS AS AN EIDETIC SCIENCE

possible — which is used in the philosophical literature to contrast the predicate 'actual'; (2) logic, mathematics, etc. are integral part of what Husserl calls 'formal ontology', that is, the science of any object whatever; and that (3), as Husserl himself says,[66]

> There is *no science of matters of fact* which, *were it fully developed as a science*, could be pure of eidetic cognitions and therefore *could be independent of the formal or the material eidetic sciences*. For, in the *first* place, it is without question that an experiential science, wherever it brings out mediate grounding of judgments, must proceed according to the *formal* principles treated by formal logic. Since, like any other science, an experiential science is directed to objects, it must be universally bound by the laws that belong to the essence of *anything objective whatever*. It thereby enters into a relation with the complex of *formal-ontological* disciplines which, besides formal logic in the narrower sense, embraces the other disciplines of "*mathesis universalis*" (for example arithmetic, pure analysis, theory of multiplicities). Moreover, in the *second* place, any matter of fact includes a *material* essential composition; and any eidetic truth belonging to the pure essences comprised in that composition must yield a law by which the given factual singularity, like any other possible singularity, is bound.

However, given that, for Husserl, essences are conceivable as abstract objects, it is legitimate to ask what sort of access does he think we have to such objects. The access is provided, in Husserl's view, by what he calls 'essential intuition' or *Wesensschau*. But what does he mean by that?

What is commonly understood by 'intuition' is the pre-reflective act of reason whereby data are presented to our consciousness. Such an act is pre-reflective in that it does not take place by means of the formulation of judgements.

Husserl distinguishes between empirical intuition, and essential intuition. For him empirical intuition allows us to identify and re-identify an object by becoming acquainted with it, that is, by representing the object in our perceptual space, rather than through the use of concepts (descriptions).[67]

On the other hand, what, according to Husserl, essential intuition presents our consciousness with is not dependent on our particular perception of a, but is something that we may also find in the perception of b, for b different from a, and which we may think about independently of our senses being engaged in the perception of the relevant a.

Mathematics is, for Husserl, a science of essences. But how are these essences given to us, since they are, after all, universals and not individuals?

---

[66][Husserl, 1998], ibid., §8, pp. 17-18.
[67][Husserl, 1998], ibid., §. 3, p. 9:

> Empirical intuition or, specifically, experience, is consciousness of an individual object; and as an intuitive consciousness it "makes this object given", as perception it makes an individual object given originarily in the consciousness of seizing upon this object "originarily", in its "personal" selfhood.

How can we have a pre-reflective representation of a universal, how can we *look at* a universal?

Suppose we have a perceptual representation (empirical intuition) of one apple or of one chair, etc. or that we, more simply, imagine or dream of one apple or one chair, etc. If we now operate on such a representation, regardless of whether this is based on perceptual data or rather on data of mere phantasy, what Husserl calls 'phenomenological reduction', i.e., if we: (1) cut out from consideration all that transcends what is directly given to us, that is, the existence of the apple or of the chair, etc. the truth of the beliefs we generate on the basis of our representations — 'There is one apple over there', 'There is one chair just here', etc.; and (2) abstract from the type of thing terms like 'apple' and 'chair' refer to and from the relations existing between the referents of these terms, we are then left with the *pure phenomenon*. This is a phenomenon which no longer presents us with a unit of a particular kind, but in which oneness is directly given to us as an object we can 'look at', to which we can refer, and about which we can make true and false statements.

At this stage of our discussion it is important to notice two things. First,[68]

> If, by some psychological miracle or other, free phantasy should lead to the imagination of data (sensuous data, for example) of an essentially novel sort such as never have occurred and never will occur in any experience, that would in no respect alter the originary givenness of the corresponding essences: though imagined Data are never actual Data.
>
> [And, therefore, the] *Positing of* and, to begin with, [the] intuitive seizing upon, *essences implies not the slightest positing of any individual factual existence; pure eidetic truths contain not the slightest assertion about matters of fact.*

Secondly, according to Husserl, oneness is *purely* immanent in the (pure) phenomenon in which it is directly given to us,[69] in the sense that the presence of oneness in the pure phenomenon is not bound to the sphere of consciousness of individuals. And, consequently, considerations relating to the necessary and sufficient psychological conditions for the acts of abstraction to take place in individuals, etc. are completely irrelevant to a discussion of essences and their being directly given to us.

However, if essences are given to us by means of essential intuition, which is a process of abstraction, let us ask ourselves, with Frege, 'how is it possible to have an (essential) intuition of a very large natural number k?' In fact, for k sufficiently large, we cannot even imagine (have a pre-reflective

---

[68][Husserl, 1998], ibid., §4, p. 11.

[69]Here the contrast between Frege's and Husserl's views could not be more stark. For, as we have seen in §3.7, for Frege numbers are not objects of intuition, nor are they ideas.

8. MATHEMATICS AS AN EIDETIC SCIENCE 109

representation of) a set from which to obtain k by abstraction. (These considerations apply *a fortiori* to transfinite cardinal numbers.)

Moreover, what happens to all those objects of mathematical investigation which are not representable in perceptual space and which we cannot even have as objects of our phantasy, but whose existence and properties are perfectly well handled by formal manipulations?

But, apart from the objections above, now someone might ask 'If mathematics is a science of essences and pure eidetic truths contain not the slightest assertion about matters of fact, how is it possible that some mathematical theories are applied to the empirical sciences?'

For Husserl, mathematical theories become applicable to the empirical sciences, i.e., to matters of fact, because, conceiving of certain individuals as singularizations of an eidetic universality,[70]

> ... eidetic universality becomes transferred to [such individuals, when we posit them] as factually existing, or to an indeterminately universal sphere of [such] individuals (which undergoes positing as factually existent).

Having established that, for Husserl, mathematics is an eidetic science and having shown that Husserl's view of mathematics has an interesting answer to give to the accessibility and applicability problems, a question springs to mind concerning what, according to Husserl, is specific of mathematics as opposed to other eidetic sciences like phenomenology itself.

For Husserl, the objects of study of mathematics are what he calls 'mathematical manifolds in the pregnant sense', where a mathematical manifold in the pregnant sense[71]

> ... is characterized by the fact that a *finite number of concepts and propositions derivable in a given case from the essence of the province in question, in the manner characteristic of purely analytic necessity completely and unambiguously determines the totality of all the possible formations belonging to the province* so that, *of essential necessity, nothing in the province is left open.*

> We can also say that such a manifold has the distinctive property of being *"mathematically-exhaustively definable"*. The "definition" consists of the system of axiomatic concepts and axioms; and the "mathematically exhaustive" consists of the fact that the defining assertions involve the greatest conceivable prejudgment [*Präjudiz*] concerning the manifold: nothing remains undetermined.

What emerges with extreme clarity from Husserl's characterization of mathematical manifolds in the pregnant sense is that their theories must be (1) finitely axiomatizable; and (2) decidable. With regard to the points just made we must notice that (finite) axiomatizability and decidability are so important for Husserl's thinking about mathematics that he places them

---

[70] [Husserl, 1998], ibid., §6, pp. 14-15.
[71] [Husserl, 1998], First Book, Part three, Chapter one, §72, p. 163.

as the necessary and sufficient conditions which must be satisfied by any deductive discipline for us to be justified in saying that it is mathematical:[72]

> A system of axioms which ... "exhaustively defines" a manifold purely analytically is what I call also *a definite system of axioms*. Any deductive discipline based on such a system is a *definite discipline* or, *in the pregnant sense*, one which is *mathematical*.

Having shed some light on what Husserl believes about the specific nature of mathematics with respect to the other eidetic sciences, it is rather surprising to realize that either or both characteristics (1) and (2) are not satisfied by most of the important mathematical theories we have.

As is well known, (a) if most of the important mathematical axiom systems we have — first-order arithmetic, **ZFC**, etc. — are consistent, they are also incomplete; and (b) incompleteness implies undecidability. From (a) and (b) follows that, if most of the important mathematical axiom systems we have are consistent, they cannot satisfy condition (2).

Secondly, there are some very important mathematical axiom systems which make use of an infinite number of axioms. In particular, **ZFC** is an axiom system, with a special recognized foundational rôle, having an infinite number of axioms which, if consistent, does not satisfy condition (2) either.

These considerations have a devastating effect on Husserl's view of mathematics. For they show that most of the important mathematical axiom systems we have do *not* exhaustively define a manifold purely analytically, to use Husserl's terminology. (As a consequence of Church's theorem, even first-order predicate logic does not satisfy condition (2)!)

In any case, independently of considerations relating to the limitations of a certain class of formal systems, the axiomatic method itself, far from being universally accepted as 'the' mathematical method, has become object of criticism. For example, against the idea that mathematical theories must be axiomatized, Brouwer observes that axiomatic systems set arbitrary and unacceptable limitations to the creative activity of the subject, creative activity which, according to Brouwer, 'develops in self-unfolding guided by free arbitrariness'[73].

Actually, for Brouwer, the axiomatic method must be rejected outright, because, among other things, it leads mathematicians astray by giving too much importance to structures and their rôle in mathematics.

According to other critics, the main defect of the axiomatic method consists, instead, in being unequal to the task of representing mathematical knowledge.[74] In particular, for Cellucci, the axiomatic method is simply a

---

[72][Husserl, 1998], ibid., §72, p. 164.
[73][Brouwer, 1907].
[74][Cellucci, 1998], Ch. VII, §8, p. 254.

very successful strategy of justification which, as a consequence of its hegemonic position within mathematics, ends up obscuring the other face of the coin: the context of mathematical discovery.

For obvious reasons of space, I am not going to discuss here the objections above,[75] which I have decided to mention only in passing so that the reader might realize that, without addressing the issues raised by them, a position such as Husserl's is bound to rest on very uncertain ground.

9 Mathematics as a science of concepts

Another Platonist attempt to answer the accessibility objection is that of Gödel. For Gödel,[76]

> Mathematical propositions ... do not express physical properties of the structures concerned [in physics], but rather properties of the *concepts* in which we describe those structures. But this only shows that the properties of those concepts are something quite as objective and independent of our choice as physical properties of matter.

Moreover, according to him, concepts, sets, and propositions form an objective reality:[77]

> What is wrong, however, is that the meaning of the terms (that is, the concepts they denote) is asserted to be something man-made and consisting merely in semantical conventions. The truth, I believe, is that these concepts form an objective reality of their own, which we cannot create or change, but only perceive and describe.

The quotations above leave little doubt concerning where Gödel's sympathies lie in the dispute between realists and anti-realists in the philosophy of mathematics, even though they seem to commit Gödel to a realist position only with regard to applicable/applied mathematics. However, the clearest passage, among the ones I know, in which Gödel professes his Platonist faith concerning, unreservedly, mathematics as a whole may be found at the end of his Gibbs lecture. There Gödel says that:[78]

> ... after sufficient clarification of the concepts in question it will be possible to conduct these discussions with mathematical rigour and ... the result then will be that (under certain assumptions which can hardly be denied [in particular the assumption that there exists at all something like mathematical knowledge]) *the Platonistic view is the only one tenable. Thereby I mean the view that mathematics describes a non-sensual reality, which exists independently both of the acts and ⟨of ⟩ the dispositions of the human mind and is only perceived, and probably perceived very incompletely, by the human mind.*[79]

---

[75]For a discussion of these issues see [Oliveri, 2005], §4.
[76][Gödel, 1953], p. 360.
[77][Gödel, 1951], p. 320.
[78][Gödel, 1951], p. 322-323.
[79]The italics are mine.

In the quotation above Gödel explicitly asserts that, for him, mathematics describes a non-sensual reality of which we have perception. But, if mathematical reality is non-sensual, how can we have perceptions of it? What does Gödel mean when he talks about 'perception by the human mind'?

It seems that, according to Gödel, we obtain knowledge of the properties of the objects studied by mathematical theories by means of a kind of intuition, which he likens to, but does not identify with, a physical sense:[80]

> The similarity between mathematical intuition and a physical sense is very striking. It is arbitrary to consider 'this is red' an immediate datum, but not so to consider the proposition expressing modus ponens or complete induction (or perhaps some simpler proposition from which the latter follows). For the difference, as far as it is relevant here, consists solely in the fact that in the first case a relationship between a concept and a particular object is perceived, while in the second case it is a relationship between concepts.

In reading the quotation above it seems to be clear that, according to Gödel, mathematical intuition, like a physical sense, produces pre-reflective representations; but that, unlike a physical sense, what is given through mathematical intuition are concepts, properties of concepts, and relations between concepts.

What, on the other hand, appears to be missing, in both the quotation above and the other philosophical writings of Gödel, is an elucidation of the notion of mathematical intuition able to justify Gödel's assertion concerning the similarity existing between mathematical intuition and sense perception.

There is, however, sufficient evidence in favour of the idea that Gödel's view of mathematical reality differs from that of Frege on the nature of the elements of what he calls the 'Third Realm'. For Frege, the entities which populate the Third Realm — thoughts — are neither ideas nor concrete objects, and *are not given to us by any kind of perception*. We grasp them through thinking, but this does not mean that they are just contents of our consciousness:[81]

> The expression 'grasp' is as metaphorical as 'content of consciousness'. The nature of language does not permit anything else. What I hold in my hand can certainly be regarded as the content of my hand; but all the same it is the content of my hand in quite another and a more extraneous way than are the bones and muscles of which the hand consists or again the tensions these undergo.

Mathematics and logic, according to Frege, have nothing to do with psychology. Gödel shares such a view with Frege, but this leaves to be explained what we are to do with the Gödelian notion of mathematical intuition. For, usually, the concept of intuition carries a strong psychological connotation,

---
[80][Gödel, 1953], p. 359.
[81][Frege 1977b,], footnote 6, pp. 24-25.

9. MATHEMATICS AS A SCIENCE OF CONCEPTS 113

which makes impossible to give a satisfactory account of mathematical objectivity: how do you know whether your pre-reflective representation of such-and-such a concept, or relation between concepts, etc. is the same as mine?

Some people have thought that, to have our cake and eat it, we could, perhaps, consider understanding as a non-psychological phenomenon and then assimilate the notion of mathematical intuition to that of understanding. This seems, indeed, to be the direction followed by Church.

According to Church, understanding is like seeing, when these two activities are considered as methods of observation:[82]

> The extreme demand for a simple prohibition of abstract entities under all circumstances perhaps arises from a desire to maintain the connection between theory and observation. But the preference of (say) *seeing* over *understanding* as a method of observation seems to me capricious. For just as an opaque body may be seen, so a concept may be understood or grasped. And the parallel between the two cases is indeed rather close. In both cases the observation is not direct but through intermediaries — light, lens of eye or optical instrument, and retina in the case of the visible body, linguistic expressions in the case of the concept. And in both cases there are or may be tenable theories according to which the entity in question, opaque body or concept, is not assumed, but only those things which would otherwise be called its effects.

It seems to me that Church's characterization of understanding is untenable and that Gödel did not have the same view as Church on these matters. To see this consider, first of all, that, for us to be justified in asserting that someone has grasped the concept of red, it is sufficient that the person we have in mind (a) is a competent speaker of English (or of Italian, etc.), and (b) uses correctly the term 'red' (or the term 'rosso', etc.). The reason for this is that, if someone uses correctly the word 'red', he has to know what must be the case if assertions like 'This is red' are true. In other words, he must be able to judge correctly whether or not, in ordinary circumstances, an object falls under the concept red. From this follows that, grasping the meaning of 'red', which entails grasping the concept of red, boils down to the ability to make a certain kind of judgment. But making judgments is a reflective activity and, therefore, it cannot be confused with any form of intuition.

Secondly, as we have already seen, Gödel speaks of 'mathematical intuition', which he likens to a physical sense, and refers to what is given through it as an 'immediate datum'. No attempt is made by Gödel to characterize intuition as a judgment making activity whereby objects are presented to our consciousness.

Another interesting suggestion concerning what Gödel thought of mathematical intuition comes from Føllesdal. According to Føllesdal, Gödel's

---

[82][Church, 1951], p. 104.

mathematical intuition resembles very closely Husserl's *Wesensschau*, intuition of essence.[83] If correct, this suggestion would satisfactorily show that Gödel's Platonist view of mathematics and his non-Fregean ideas concerning the access we have to the realm of abstract mathematical objects are fully compatible with one another.

Although in Gödel's writings there does not seem to be an explicit endorsement of phenomenology, Gödel certainly had first-hand knowledge of, and a great appreciation for, Husserl's philosophy:[84]

> I believe that precisely because in the last analysis the Kantian philosophy rests on the idea of phenomenology, albeit in a not entirely clear way, and has just thereby introduced into our thought something completely new, and indeed characteristic of every genuine philosophy — it is precisely on that, I believe, that the enormous influence which Kant has exercised over the entire subsequent development of philosophy rests. Indeed, there is hardly any later direction that is not somehow related to Kant's ideas. On the other hand, however, just because of the lack of clarity and the literal incorrectness of many of Kant's formulations, quite divergent directions have developed out of Kant's thought — none of which, however, really did justice to the core of Kant's thought. This requirement seems to me to be met for the first time by phenomenology, which, entirely as intended by Kant, avoids both the death-defying leaps of idealism into a new metaphysics as well as the positivistic rejection of all metaphysics. But now, if the misunderstood Kant has already led to so much that is interesting in philosophy, and also indirectly in science, how much more can we expect from Kant understood correctly?

As a matter of caution, we must notice that, if Føllesdal is right, and Gödel had, in fact, adopted a phenomenological standpoint concerning mathematics, this must have, in any case, differed somewhat from Husserl's own. For, as we have already seen in §3.8, Gödel's Incompleteness Theorems make it impossible to believe, as Husserl did, that all mathematical theories are finitely axiomatizable and decidable.

However things might be with regard to this question, it is important, before bringing this section to a close, to dwell a little longer on the concept of mathematical intuition, which, as we have already seen, performs a very central rôle in both Husserl's and Gödel's philosophies of mathematics.

One of the scholars who, prompted by his interest in the study of Gödel's philosophy of mathematics, has contributed to the elucidation of this concept is C. Parsons.

For Parsons, if mathematical intuition[85]

> ...is to be central to the philosophy of mathematics, it should play a role [concerning mathematical objects, and perhaps other abstract entities] like that of sense perception in our knowledge of the everyday world and of physics.

---

[83]See on this [Føllesdal 1999,].
[84][Gödel, 1961], pp. 385 and 387.
[85][Parsons, 1980], p. 95.

9. MATHEMATICS AS A SCIENCE OF CONCEPTS   115

According to Parsons, mathematical intuition is a particular example of a much broader representational ability, which appears to 'play a rôle like sense perception, in our knowledge of the everyday world and of physics'. Parsons calls this representational ability 'intuition of types'.

For Parsons, the intuition of types is based on our perception of tokens. To understand what he means, imagine[86]

> ...a 'language' with a single basic symbol '|' (stroke), whose well-formed expressions are just arbitrary strings containing just this symbol, that is, |, ||, |||, ...This sequence of strings is isomorphic to the natural numbers, if one takes '|' as 0 and the operation of adding one more '|' on the right as the successor operation. This yields an interpretation of arithmetic as a kind of geometry of strings of strokes. At first sight the interpretation leaves out the concept of *number*, that is the role of natural numbers as cardinals and ordinals.
>
> Ordinary perception of a string of strokes would have to be perception of a *token*, but we naturally think of such symbols as types.

To spell out the point just made, and show that the scope of the intuition of types includes, and is much broader than, that of mathematical intuition, Parsons adds that[87]

> ...the widespread impression that mathematical intuition is a 'special' faculty, which perhaps comes into play only in doing pure mathematics

can be dispelled, if we consider that[88]

> ...in some cases, taking what is given as a type is quite spontaneous and natural. The most obvious is the understanding of natural language: the hearer is without reflection ready to re-identify the type (in the linguistic, not the acoustic, sense). Typically, the hearer of an utterance has a more explicit conception of *what was uttered* (e.g., what words) than he has of an objective identification of the *event* of the utterance. I believe that the same is true of some other kinds of universals, such as sense qualities and shapes. Indeed, in all these cases it seems not to violate ordinary language to talk of perception of the universal *as an object*, where an instance of it is present.

But, according to Parsons, there are three main disanalogies between mathematical intuition and sense perception, which appear to undermine the resemblance between these two representational abilities.

First, whereas, in ordinary cases of sense-perception, there exists a causal chain linking the object perceived with our perception of it, it seems '...implausible to suppose that in *mathematical* intuition there is a causal action of a mathematical object on us (presumably on the mind)'.[89]

Secondly, the objects perceivable with our senses are individually identifiable, but, since many mathematical entities like the natural numbers, the

---
[86] [Parsons, 1980], §IV, p. 103.
[87] [Parsons, 1980], ibid., p. 104.
[88] [Parsons, 1980], ibid.
[89] [Parsons, 1980], §II, p. 98.

reals, etc. can be characterized only up to isomorphism, this does not seem to be possible with mathematical objects.

Thirdly, in mathematical intuition '...we do not have the scope that we have with ordinary perception for identifying the object of intuition independently of the subject's conceptual resources'.[90]

Concerning the first disanalogy, it is important to notice that, if the argument supporting it is: if mathematical objects are abstract objects, and abstract objects are not causally efficacious, it follows that mathematical objects exercise no causal action on us; the assertion of the first disanalogy is unwarranted. For, since one of the premisses of the argument above is false, the argument is not sound.

To see that it is false to assert that abstract objects are not causally efficacious, consider that, in our daily life and exchanges with other people, the meanings of flashing traffic lights, things people say, etc. cause us to act in particular ways rather than others, and that meanings are abstract entities.

Moreover, for Husserl, the first disanalogy would not be a worry because of its irrelevance. In fact, since in the context determined by phenomenological reduction, it is nonsensical to appeal to causality in general, it is nonsensical, in particular, to use the concept of causality to tell apart anything from anything else.

The second disanalogy can be tackled by means of a structuralist approach to mathematics and empirical science. If mathematics is not really about objects, but is about structures (see §3.10), and if, on the basis of Quinean arguments relying on inscrutability of reference, we have structures not only in mathematics, but 'all the way down' also in the empirical sciences,[91] vanishes the illusion of being able to identify individually the objects perceivable with our senses, perhaps by pointing at them.

With regard to the third disanalogy, we can readily concede that, in the case of mathematics, we do not go anywhere without 'the subject's conceptual resources', and that there are cases of 'ordinary perception' in which we are able to identify and re-identify an object simply by being acquainted with it.

But, on the other hand, Parson's requirement that, to be central to the philosophy of mathematics, mathematical intuition should play a rôle like that of sense perception in our knowledge of the everyday world and of physics implies that, if by 'knowledge of the everyday world' we mean the information produced by physics and experimental psychology concerning the existence and properties of entities we have perception of without the

---

[90][Parsons, 1980], §VII, p. 110.
[91]See on this [Quine, 1981a].

aid of instruments of observation, then we should conclude that conceptual content is an essential component of sense perception. This, of course, implies that, even when sense perception is involved with our knowledge of the everyday world and of physics, we cannot identify the object of intuition independently of the subject's conceptual resources.

The discussions of this section make one hope in the existence of a viable concept of mathematical intuition which could go well together with some Platonist view of mathematics as a science of structures.

10 Platonism about structures

Having examined three different ways of defending Platonism about objects through providing an answer to the accessibility problem, time has now come for an examination of the second Platonist strategy implemented against the anti-realist's objections to the viability of Platonism about objects: Platonism about structures. But before I do so, I shall briefly consider yet another proposed defense of Platonism about objects.

According to Balaguer the non-uniqueness of natural numbers (and other mathematical entities which can be characterized only up to isomorphism) is not a problem for the Platonist. In fact, for Balaguer, the Platonist can embrace what he calls 'full-blooded Platonism' (**FBP**), that is,[92]

> ...the view that all the mathematical objects that (logically) possibly *could* exist actually *do* exist

relinquishing at the same time the traditional Platonist thesis according to which

> (**P**) '...there *is* a unique sequence of abstract objects which is the natural numbers.'[93]

It is obvious how giving up (**P**) removes the sting from the anti-Platonist argument based on the existence of different isomorphic models of number theory (and of other mathematical theories).

However, with regard to what has just been stated, it is important to notice that, according to the believers in **FBP**, a 'mathematical object which *could* exist' is an object studied/postulated by a consistent mathematical theory.

Therefore, for the full-blooded Platonist[94]

> ...every consistent purely mathematical theory truly describes some collection of abstract mathematical objects. Thus, to acquire knowledge of abstract objects,

---

[92][Balaguer, 1998], §4, p. 75.
[93][Balaguer, 1998], §1, p. 64.
[94][Balaguer, 1998], §4, p. 75.

all we need to do is acquire knowledge that some purely mathematical theory is consistent.

According to Balaguer, **FBP**, besides dissolving the non-uniqueness problem, would also be immune to the accessibility problem, because having[95]

> ...knowledge of the consistency of a mathematical theory — or any other kind of theory for that matter — does not require any sort of *access* to the objects that the theory is about.

In considering Balaguer's proposal, it seems to me that the reliance of **FBP** on the consistency of mathematical theories — both to formulate the concept of existence of mathematical objects in terms of logical possibility, and to circumvent the accessibility problem — might be affected by a fatal fault.

In fact, since, according to **FBP**, by a 'mathematical object which *could* exist', we must mean a mathematical object studied/postulated by a consistent mathematical theory, we have that, as a consequence of Gödel's Incompleteness Theorems, it is in general impossible to make any use of such a notion. (An argument drawn along similar lines shows that **FBP** is not immune to the accessibility problem either.)

Given the assessment above of an extreme attempt to salvage Platonism about objects, we can now resume the route leading to a discussion of Platonism about structures.

The analysis produced in what follows will reveal that the position in the philosophy of mathematics I have called 'Platonism about structures' suffers from very serious defects connected with the underlying Platonism and not with the structuralist view of mathematics.

Responding to the anti-Platonist's argument based on the existence of different isomorphic models of a given consistent axiomatic system $\mathcal{G}$, the Platonist can say that $\mathcal{G}$ describes the properties of what makes all these models of $\mathcal{G}$ isomorphic to each other, that is, what $\mathcal{G}$ describes is a certain abstract set of objects $\mathcal{O}$ and a set of relations $\mathcal{R}$ defined on $\mathcal{O}$. Let us call the pair $(\mathcal{O}, \mathcal{R})$ a 'structure'.

When questioned about the nature of arithmetic, and of mathematics in general, the Platonist can, therefore, reply with the well known Quinean quip to the effect that '...arithmetic is all there is to number',[96] and add that mathematics in general is a science of structures $(\mathcal{O}, \mathcal{R})$ — Resnik calls them 'patterns' — in which one abstracts from the nature of the individual objects belonging to the domain $\mathcal{O}$. And when pressed further for an explanation of what a structure is, he can expand on the characterization given

---

[95] [Balaguer, 1998], ibid.
[96] [Quine, 1969], §I, p. 45.

11. PLATONISM ABOUT STRUCTURES, ITS IMPORTANCE

above saying that a structure is:[97]

> ...a complex entity consisting of one or more objects, [called] *positions*, standing in various relationships (and having various characteristics, distinguished positions and operations).

However, in talking about mathematical structures we must distinguish between two different ways of conceiving them. The first way is as *ante rem* entities, that is, as entities which exist prior to (independently of) the systems (sets of objects and relations defined on such sets) which realize them.

The second way of thinking of them is as entities which exist *in re*, that is, as entities which are realized by given systems, and do not exist independently of such systems.

The genuinely Platonist view of structures is the first, and it presently has among its staunchest supporters M. Resnik and S. Shapiro. The second view of structures is more Aristotelian in character, and is what captures the sympathies of Quine.

11 On the philosophical importance of Platonism about structures

Platonism about structures presents several features which make it very attractive to someone who is realist about mathematics.

First, Platonism about structures is able to transform a weak point of Platonism about objects into a strength of its own position. Indeed, as we have seen in the previous section, the anti-Platonist argument from isomorphic models of a consistent axiomatic system $\mathcal{G}$ can be transformed into a powerful argument in favour of the idea that mathematics is a science of structures.

Secondly, since the Platonist about structures has an *ante rem* view of structures, and a structure is nothing but a set of objects and relations defined on this set, he can also be a Platonist about objects.

In fact, the Platonist about structures can believe in the existence of mathematical objects, say, the natural numbers, with the proviso that the characterization of such entities is given in terms of the elements of an abstract progression, an abstract progression which we obtain, starting from an ordinary progression, by abstraction from the nature of the elements of the ordinary progression.[98]

---

[97][Resnik, 1981], p. 532.
[98]Notice that the Platonist about structures does not have to deny that two different elements of the abstract progression *intrinsically differ* from one another. He can simply consider mathematically irrelevant to determine what sort of difference this is.

It is interesting to notice that also an *in re* mathematical structuralist *à la* Quine can believe in mathematical objects. Indeed, according to Quine:[99]

> ...any progression will serve as a version of number so long and only so long as we stick to one and the same progression. Arithmetic is, in this sense, all there is to number: there is no saying absolutely what the numbers are; there is only arithmetic.

We must realize that both the positions above which, as it were, reinstate the dignity of mathematical objects within a structuralist view of mathematics, differ from Resnik's.

For, according to Resnik,[100]

> That 13 is a prime number is not determined by some internal property of 13 but rather by its place in the structure of the natural numbers.

In other words, for Resnik, there is no more to mathematical objects than what is determined by the relations of the structure to which they belong.

Now, it is important to notice that Resnik's view is untenable, because if we apply it to the following situation:

$$P: \quad 0, \quad 1, \quad 2, \quad \ldots$$
$$P': \quad 1, \quad 2, \quad 3, \quad \ldots$$
$$P'': \quad 2, \quad 3, \quad 4, \quad \ldots$$
$$\vdots \quad \vdots \quad \vdots \quad \vdots$$

we should deduce that

$$0 = 1, \; 0 = 2, \; 0 = 3, \ldots$$

But this is absurd, because it would contradict our hypothesis concerning the fact that $P, P', P'', \ldots$ are isomorphic models of Peano Arithmetic (**PA**) which differ from one another.

Thirdly, Platonism about structures seems to agree with the development of the history of mathematics, because it captures the trend of evolution of mathematics since the nineteenth century. The present day study of many different branches of mathematics such as algebra and topology can be correctly described as a study of structures.

Fourthly, if Quine's arguments in favour of inscrutability of reference are correct, and therefore also empirical science has to be seen as a science of structure, Platonism about structures in mathematics would have a very plausible, and entirely general, way of explaining the applicability of mathematics to science which proceeds from the very nature of mathematical

---

[99][Quine, 1969], §I, p. 45.
[100][Resnik, 1981], p. 529.

12. PLATONISM ABOUT STRUCTURES: SOME OBJECTIONS

investigation. Such an explanation would consist in the consideration that the structures studied by mathematics have such a level of generality that their study is applicable to a vast number of less general structures such as those realized by 'concrete systems', that is, by sets of concrete objects and actual relations defined on them. It is clear how this would be a complete vindication of the position of '...those who think, like Frege, that the claim of mathematics to be a science derives from its applications...'[101]

But if, as we have seen in this section, Platonism about structures presents several advantages over Platonism about objects, what are its weak points?

12 Some objections to Platonism about structures

Platonism about structures has been, for some time, the target of many objections. Russell was one of the first philosophers to be sceptical about a structuralist approach to arithmetic. I shall here mention two of his perplexities together with Quine's replies.

For Russell, if we want to obtain[102]

> ...an understanding of number the laws of arithmetic are not enough; we must know the applications, we must understand numerical discourse embedded in discourse of other matters. In applying number, the key notion, he urged, is *Anzahl*: there are n so-and-sos. However, Russell can be answered. First take, specifically, *Anzahl*. We can define "there are n so-and-sos" without ever deciding what numbers are, apart from their fulfillment of arithmetic. That there are n so-and-sos can be explained simply as meaning that the so-and-sos are in one-to-one correspondence with the numbers up to n.

> Russell's more general point about application can be answered too. Always, if the structure is there, the applications will fall into place.

But, assuming that we can talk about structures in terms of abstract entities which exist independently of anybody's thinking of them, for a believer in the Quinean dictum 'no entity without identity', it is legitimate to ask what sort of identity criterion we have for them. One of the most influential answers to this question is that given by Michael Resnik.

For Resnik, patterns (structures) are the objects of description of consistent mathematical theories or, as he puts it, of formal systems. Given a consistent mathematical theory $\mathcal{T}$, a model $\mathcal{M}$ of $\mathcal{T}$ would be, for him, an instantiation of the pattern described by $\mathcal{T}$.[103]

---

[101] [Dummett, 1995], Ch. 23, p. 297.
[102] [Quine, 1969], §I, p. 44.
[103] [Resnik, 1981], pp. 532–533:

> Thinking of patterns as models of formal systems, ... [congruence] is the relationship which holds between *isomorphic models* of formal systems. ... When a pattern and an arrangement of so-called concrete objects such as ... puppies are congruent then I say that the arrangement *instantiates* the pattern. ... [Ten puppies in a litter] thus instantiate the one-to-ten pattern.

This Platonist position on patterns, as *ante rem* structures instantiated by the models of the theories which describe them, allows Resnik to formulate identity criteria for structures. According to Resnik, two structures are identical if and only if they are instantiated by models which are isomorphic to one another.[104]

However, the solution chosen by Resnik to the problem of providing identity conditions for structures gets him into trouble. In fact, now the anti-Platonist can object that Resnik's view of patterns/structures is not general enough to provide a justification for saying that mathematics is a science of patterns, because it only applies to structures described by so-called 'categorical axiom systems'. And this is a family from the membership of which are excluded many very important mathematical axiom systems like those of group theory, topology, set theory, etc.

Moreover, the anti-Platonist would continue, we are justified in saying that non-categorical axiom systems describe structures, because (1) what they describe falls under the definition of structure given in §3.10 — a definition widely accepted within the mathematical community; (2) non-categorical axiom systems like those of group theory, topology, set theory, etc. provide us with identity conditions for the structures they describe in the sense that: (a) they give us the opportunity to find out whether something is a group or not, a topological space or not, a set or not, etc.; and (b) if G is a group, etc. they also supply us with identity criteria for G (group-isomorphism), etc.

Replying to this objection saying that 'Since the main proponents of the structuralist approach — Resnik and Shapiro — hold that structures are defined by those axiom systems that *are* categorical, it's hard to see what the point is of noting the existence of non-categorical axiom systems' would be most unsatisfactory. For, it would seem to suggest that the correctness of a definition in mathematics depends on someone's whim rather than on mathematical practice and reality. Holding this view would be as ludicrous as saying that, since someone had a long time ago decided that by 'number' we must mean the positive integers and the ratios of positive integers, 'it's hard to see what the point is of noting the existence of 0, the negative integers, etc. to show that the already mentioned definition of number is inadequate'.

But, returning to the problem of what ought to be considered to be a mathematical structure, it is interesting to notice how clear the position of

---

[104][Resnik, 2001], Ch. 10, §3, p. 209:
> ...we identify patterns just in case they have the same positions and the same relations...

12. PLATONISM ABOUT STRUCTURES: SOME OBJECTIONS 123

the mathematical community is on this point:[105]

> The first axiomatic treatments and those which caused the greatest stir (those of arithmetic by Dedekind and Peano, those of Euclidean geometry by Hilbert) dealt with univalent theories, i.e., theories which are entirely determined by their complete system of axioms; for this reason they could not be applied to any theory except the one from which they had been extracted (quite contrary to what we have seen, for instance, for the theory of groups). If the same had been true for all other structures, the reproach of sterility brought against the axiomatic method would have been fully justified. But the further development of the method has revealed its power; and the repugnance which it still meets here and there, can only be explained by the natural difficulty of the mind to admit, in dealing with a concrete problem, that a form of intuition, which is not suggested directly by the given elements (and which often can be arrived at only by a higher and frequently difficult stage of abstraction), can turn out to be equally fruitful.

However, independently of the discussion concerning the relative merits of various forms of Platonism about structures, it must be remarked that this philosophy of mathematics, like Platonism about objects, has the accessibility problem to solve. The fact that structures are thought to be abstract *ante rem* entities poses very strongly the problem of how we come to know about the existence and properties of such things.

According to Resnikian Platonism about structures (**RPS**), there are three ways in which we access structures. The first is abstraction or pattern recognition:[106]

> We do not literally see or hear abstract entities. Our subject comes to recognize a pattern by observing patterned systems.

The second is linguistic abstraction:[107]

> ... humans have a faculty that resembles pattern recognition but goes beyond simple abstraction. The small finite structures, once abstracted, are seen to display a pattern themselves. For example, the finite cardinal structures come in a natural order: the 1 pattern, followed by the 2 pattern, followed by the 3 pattern, and so on. We then *project* this pattern of patterns beyond the structures obtained by simple abstraction.

The third is implicit definition:[108]

> At its root ... an implicit definition is a collection of sentences, which we can call "axioms". Of course, not every set of sentences successfully characterizes a structure, even if someone intends to use it for that purpose ... There are two requirements on an implicit definition. The first is that *at least* one structure satisfies the axioms. Call this the "existence condition". The second requirement is that *at most* one structure (up to isomorphism) is described. This is the "uniqueness condition".

---

[105][Bourbaki, 1948], §7, p. 1275.
[106][Shapiro, 1997], Ch. 4, §1, p. 111.
[107][Shapiro, 1997], ibid., §3, p. 118.
[108][Shapiro, 1997], ibid., §8, p. 132.

Let us now consider these three points in turn. The first of *the three ways* of accessing structures, according to **RPS**, consists in 'recognizing a pattern by observing a patterned system'. To reinforce the plausibility of the above point, we are told by the **RPS** believer that even chickens, who do not have a reputation for being among the most astute representatives of the animal kingdom, can recognize patterns by observing patterned systems. The question, though, is not about whether it is possible, given a patterned system, to recognize some pattern in it — imagine some of Pavlov's experiments of pattern recognition based on conditioned reflexes — but, given what the nature of mathematical structures is for **RPS**, whether *mathematical* patterns (structures) are so recognizable.

Moreover, if, as Resnik seems to hold,[109]

> Seeing a pattern is more a matter of *seeing that* certain of its instances fit it or satisfy its defining conditions [and since] To abstract a pattern from instances is neither to intuit nor to see it; rather it is a process by which we arrive at a description of the pattern by alternatively positing related positions and checking their fit against putative instances

it is legitimate to ask whether this process can be correctly described as one of abstracting a pattern from instances.

Indeed, if abstracting a pattern from instances presupposes that the pattern has already been defined, if only implicitly, by means of a set of axioms; and consists in checking that (*seeing that*) 'certain of its instances fit it or satisfy its defining conditions', it seems safe to conclude that such a process is more similar to the act of testing a (set of) conjecture(s) *imposed upon* patterned systems (instances) than to that of abstracting a pattern from patterned systems.

The second way of accessing abstract structures, according to **RPS**, evokes a pleasing picture. We first access *small* abstract structures by pattern recognition (or simple abstraction), and having, for instance, accessed 'the 1 pattern, followed by the 2 pattern, followed by the 3 pattern, and so on. We then *project* this pattern of patterns beyond the structures obtained by simple abstraction.'

One of the perplexing features of this picture is the idea of projecting a pattern of patterns. In fact, how can we possibly say that such an act of projection is a way of accessing an abstract mathematical structure which is given *ante rem* rather than simply being a creative act similar to that of writing (projecting) a fiction story?

Moreover, the use of the term 'projection' here seems to be evoking the act of extending or completing something, but, if this is the case, how can we possibly think of our mathematical structures as existing independently of

---

[109][Resnik, 2001], Ch. 11, §1, p. 225.

12. PLATONISM ABOUT STRUCTURES: SOME OBJECTIONS   125

such an act? If we cannot, we would be giving up not only the possibility of justifying our *ante rem* view of *large* structures, but realism about structures altogether.

Lastly, we have to ask ourselves whether the notion of implicit definition, enriched with the existence and uniqueness conditions imposed by Resnik and Shapiro upon what satisfies it, provides us with a way of accessing abstract structures. The answer to this question has to be 'No', because the uniqueness and existence conditions imposed upon what satisfies implicit definitions are not sufficient to explain *how* a set of sentences can give us access to an abstract structure.

CHAPTER 4

ANTI-REALISM IN MATHEMATICS

1 Introduction

In Chapter 3 we noticed that the term 'realism', when used within the philosophy of mathematics, rather than denoting one particular metaphysical view, refers to a wide spectrum of positions. Such positions share the idea that mathematics is a science that describes existing entities, but differ from one another on what sort of entities these are, on how we come to know them and their properties, etc.

As we shall see in this chapter, the use of the term 'anti-realism' within the philosophy of mathematics does not refer to a particular doctrine either, but covers, as it were, a multitude of *Sinn*(s). The factor common to all the metaphysical doctrines which fall under the blanket-term 'anti-realism' is that mathematics is *not* a science that describes existing entities.

Anti-realism in the philosophy of mathematics originates from the idea that it is impossible for the realist to provide a satisfactory epistemology which would harmonize with and, therefore, substantiate his view according to which mathematics produces descriptions of existing entities.

In what follows I shall, among other things, argue that, although anti-realism is motivated by a legitimate worry about the grounds on which the realist intends to justify his position, it is not more plausible than realism as a metaphysical basis for a satisfactory philosophy of mathematics.

2 Formalism and Hilbert's programme

According to formalism:[1]

> ... [a mathematical theory] is a game with signs, which are called empty. That means they have no other content (in the calculating game) than they are assigned by their behaviour with respect to certain rules of combination (rules of the game).

Formalism is, clearly, an anti-realist philosophy of mathematics. In fact, if mathematical theories are games, they must be seen, on a par with chess, draughts, etc., as the outcome of the creative, and not descriptive, activity

---

[1][Frege, 1977], §88, p. 183.

of mankind. And, in spite of the several attacks of which it has been the object, especially at the hands of Platonist philosophers like Frege and Gödel, formalism has remained — as Ruben Hersh has shown — the view that a large part of contemporary working mathematicians have of their subject.[2]

In the literature formalism is often conflated with Hilbert's programme. But that Hilbert was no formalist is shown by the following quotation:[3]

> ... we find ourselves in agreement with the philosophers, notably with Kant. Kant taught — and it is an integral part of his doctrine — that mathematics treats a subject matter which is given independently of logic. Mathematics, therefore, can never be grounded solely on logic. Consequently, Frege's and Dedekind's attempts to so ground it were doomed to failure.
>
> As a further precondition for using logical deduction and carrying out logical operations, something must be given in conception, viz., certain extralogical concrete objects which are intuited as directly experienced prior to all thinking. For logical deduction to be certain, we must be able to see every aspect of these objects, and their properties, differences, sequences, and contiguities must be given, together with the objects themselves, as something which cannot be reduced to something else and which requires no reduction. This is the basic philosophy which I find necessary, not just for mathematics, but for all scientific thinking, understanding, and communicating.

For Hilbert,[4]

> The subject matter of mathematics is ... the concrete symbols themselves whose structure is immediately clear and recognizable.

To understand what he means by this, and how different Hilbert's view of mathematics is from that of the formalist, it is useful to consider an example taken from number theory and discussed by Hilbert himself:[5]

> In number theory we have the numerical symbols
>
> $$1, 11, 111, 1111$$
>
> where each numerical symbol is intuitively recognizable by the fact that it contains only 1's. These numerical symbols which are themselves our subject matter have no significance in themselves. But we require in addition to these symbols, even in elementary number theory, other symbols which have meaning and which serve to facilitate communication; for example the symbol 2 is used as an abbreviation for the numerical symbol 11, and the numerical symbol 3 as an abbreviation for the numerical symbol 111. Moreover, we use symbols like $+, =,$ and $>$ to communicate statements. $2 + 3 = 3 + 2$ is intended to communicate the fact that $2 + 3$ and $3 + 2$, when abbreviations are taken into account, are the selfsame numerical symbol, viz., the numerical symbol 11111. Similarly $3 > 2$ serves to communicate the fact that the symbol 3, i.e., 111, is longer than the symbol 2, i.e., 11; or, in other words, that the latter symbol is a proper part of the former.

---

[2] See on this [Dales, 1998].
[3] [Hilbert, 1926], pp. 191-192.
[4] [Hilbert, 1926], p. 192.
[5] [Hilbert, 1926], pp. 192–193.

2. FORMALISM AND HILBERT'S PROGRAMME

> We also use letters $\mathfrak{a}, \mathfrak{b}, \mathfrak{c}$ for communication. Thus $\mathfrak{b} > \mathfrak{a}$ communicates the fact that the numerical symbol $\mathfrak{b}$ is longer than the numerical symbol $\mathfrak{a}$. From this point of view, $\mathfrak{a} + \mathfrak{b} = \mathfrak{b} + \mathfrak{a}$ communicates only that the numerical symbol $\mathfrak{a} + \mathfrak{b}$ is the same as $\mathfrak{b} + \mathfrak{a}$. The content of this communication can also be proved through material deduction. Indeed, this kind of intuitive material treatment can take us quite far.

From the quotation above we have that numerical expressions like '2', '3', etc. are not meaningless pieces of a game which has to be played according to arbitrary syntactic rules. On the contrary, '2', '3', etc. are clearly characterized as denoting terms which refer, respectively, to '11' and '111', etc. Moreover, also the function symbol '+' is intended as referring to the operation of concatenating two elements of what we are going to call the 'basic sequence', i.e.,

$$1, 11, 111, 1111, \ldots$$

From this follows that, for Hilbert, number theory is not a game, but a science which gives us information concerning the existence and properties of the elements of the basic sequence.

What has led to misunderstanding Hilbert's view of mathematics as a variety of formalism is the very important rôle performed in Hilbert's programme by the formalization of a mathematical theory T.

Indeed, according to Hilbert, the process of formalization of a mathematical theory T^6 has nothing to do with the idea of showing that mathematical theories are games, or that they can be treated as games.

For Hilbert, the only aim of the formalization of a mathematical theory T is that of reducing T^7

> ...to form and rule [so that] There is no longer ambiguity about what constitutes a statement of the theory, or what constitutes a proof in the theory.

This is, for him, the preliminary step that must be taken in the construction of a proof-theory whose aim is to show that no contradiction can be derived within the formalized theory.

---

[6] For Hilbert, if we want to formalize a mathematical theory T, we must: (1) axiomatize T in such a way that all properties and undefined terms of T which are relevant to the proof of theorems of T are expressed by axioms; (2) translate the expressions of the language of T into expressions of another language in such a way as to preserve the logical form of the expressions of T leaving out the meaning of the technical terms occurring in the expressions; (3) apply the same process described by (1) and (2) to the logical principles of inference used in the proofs of theorems of T. What we obtain as a result of (1)–(3) is F_T, the formalization of T.

[7] [Kleene, 1974], Part I, Ch. I, §15, p. 63.

3 Formalism: a critical assessment

Several are the attractive features of formalism. Within formalism there is a pre-established harmony between ontology and epistemology, in the sense that being there nothing that is investigated by mathematical theories, there is no such a thing as mathematical knowledge (of reality) either.

Secondly, formalism poses no challenge to classical mathematics. On the contrary, it seems to be a philosophy of mathematics which receives confirmation from the central rôle that the axiomatic method has within contemporary mathematics.[8]

Thirdly, formalism has a liberating effect from the fetters that all too often traditional intuitions concerning the meaning and importance of certain mathematical concepts, conjectures, results, proof techniques, etc. put on the development of mathematical theories.

Indeed, for the formalist, developing formal systems of non-Euclidean geometries, finite geometries, infinitely dimensional vector spaces, paraconsistent logics, ZFC + ¬CH, non-commutative algebras, etc. is as legitimate a mathematical activity as studying 2- and 3-dimensional Euclidean geometry, classical logic, ZFC + CH, commutative algebras, etc.

On the other hand, formalism suffers from a number of very serious defects which not only counterbalance its positive features, but make it impossible for it to be a viable philosophy of mathematics. In what follows I shall discuss only some such defects.

First of all, the formalist, in his approach to mathematics, finds himself before a dilemma: either he must renounce 'global formalism', that is, the view according to which all mathematical theories are formal games, or his philosophy of mathematics can receive a head on refutation from mathematical practice.

The first horn of the dilemma, i.e., renouncing global formalism, is not acceptable for the formalist, if formalism has to be a thesis about the nature of mathematics. On the other hand, the second horn of the dilemma is clearly not acceptable for him either.

But is it really the case that a head on refutation of global formalism can be produced? As Frege remarked in the *Grundgesetze*, given any formal system $\mathcal{F}$, we have a whole host of genuine mathematical questions generated *about* $\mathcal{F}$, e.g., 'Is $\mathcal{F}$ consistent?', 'Is $\mathcal{F}$ complete?', etc. which are the object of study of $M(\mathcal{F})$, the metatheory of $\mathcal{F}$.

Now, since $M(\mathcal{F})$ is *de facto* a legitimate mathematical theory, and it

---

[8] As is well known, the application of the axiomatic method to a mathematical theory T consists in showing that the theorems of T are provable from a set A of unproved statements of T such that A is consistent, recursive, and the statements of A are independent of one another.

3. FORMALISM: A CRITICAL ASSESSMENT

cannot be recognized as such by the formalist, because $M(\mathcal{F})$ has meaning, it follows that global formalism is untenable as a philosophy of mathematics.

At this point the global formalist may respond to the objection above saying that we have been too quick in reaching our conclusion, because $M(\mathcal{F})$, like any other mathematical theory, can be formalized. However, such a counter objection is not effective, because the question relevant to the assessment of global formalism as a tenable philosophy of mathematics is not 'Once $M(\mathcal{F})$ is given, can $M(\mathcal{F})$ be simulated by a formal game plus a suitable interpretation?', but rather 'Is $M(\mathcal{F})$ a formal game?'.

Of course, the answer to the latter question can only come from a study of the problems from which $M(\mathcal{F})$ originated, and when we engage in such an investigation, we discover that these problems are all connected with the study of some features of $\mathcal{F}$. Therefore, $M(\mathcal{F})$ is meaningful and, consequently, global formalism is refuted.

Secondly, as is well known, Gödel's first incompleteness theorem shows that, a very large class of important mathematical theories cannot be fully axiomatized. But, of course, since limitations which cannot be overcome in the axiomatization of a mathematical theory $\mathcal{F}$, are also limitations which cannot be overcome in the process of formalization of $\mathcal{F}$, it follows that global formalism cannot be right.

Thirdly, since, for the formalist, reference and intuition have nothing to do with mathematics, a proof of the absolute consistency of formal systems becomes a criterion of paramount importance for the justification of the acceptance of formal systems.

But if, as a consequence of Gödel's incompleteness theorems, no such a proof can be produced for a very large class of important formal systems, it follows that the acceptance of a formal system belonging to this class is bound to be arbitrary.

Fourthly, taking for granted that, in spite of the limitations exposed by Gödel's incompleteness theorems, the axiomatic method still remains the royal road which gives us access to contemporary mainstream mathematics, we must observe that its connection with formalism, as its philosophy, must not be taken for granted.

In fact, if we study the process according to which axioms are chosen, we realize that convention and arbitrariness have very little to do with it. (See §2.4.)

Indeed, the history of mathematics shows that the formulation of formal (axiomatic) systems, and their extensions, often takes place on the basis of pre-existing informal, contentful theories, as an attempt to streamline them, eliminate paradoxes, characterize new objects of investigation, etc.

Arithmetic, for example, had existed for thousands of years before Peano and Dedekind axiomatized it, and, in its 'informal stage', arithmetic had existed as a theory in which the assumptions that were eventually codified in what we now know as the Peano axioms provided content to the primitive notions of zero, successor and natural number. But, of course, the process of axiomatization of number theory did not get rid of the content of informal number theory. It, actually, provided an implicit definition of zero, successor and natural number.

Certainly, once an informal mathematical theory is axiomatized, one often discovers that its axioms are not satisfied only by the so-called 'intended interpretation', but that there also are other possible interpretations which satisfy them.

However, the existence of a plurality of interpretations of an axiom system cannot be seen as what shows that the axiomatized theory is a content-free formal game, but is rather evidence in favour of the idea that what is described by the axioms of the theory is a mathematical structure.

What this means is that, when we study it within the framework provided by the history of mathematics, the process of axiomatization of an informal, contentful, mathematical theory, far from deleting the content of the informal theory, appears, instead, to preserve it making such a content explicit and clearly laid out. And, it is such a process of making the content of an informal mathematical theory explicit and clearly laid out what contributes to the emergence of the true object of investigation of mathematical theories: mathematical structure.

Now, taking for granted that the argument above is compelling, we realize that one of the assumptions on which it rests is the idea that informal mathematical theories are contentful. But why should it be so?

In each mathematical theory we can distinguish between two different types of notion. The first type is represented by notions which are defined in terms of other notions belonging to the theory, whereas the second by the so-called 'primitive notions', i.e., notions the meaning of which is not given in terms of other notions belonging to the theory.

The distinction between primitive and definable notions of a mathematical theory is crucial to provide an explanation of the reason why informal mathematical theories have content.

In fact, whereas through the process of axiomatization of a mathematical theory it is possible to give an implicit definition of the primitive notions of the theory using to this end some of the axioms, in the case of an informal mathematical theory, it is not possible even to attempt to consider the theory as a deductive engine powered by syntax alone, because the meaning of the primitive notions of the theory cannot be reduced to, or given by, 'syntactic expressions/rules' of the theory.

3. FORMALISM: A CRITICAL ASSESSMENT

Lastly, if, as argued in Chapter 2, mathematics is indispensable for the development of some empirical sciences, it follows that formalism is untenable as a philosophy of mathematics. For, if we are holist about the theories which belong to the empirical sciences, it follows that the testable success of one such a theory for which mathematics is indispensable would confirm the correctness of our commitment to the things the theory says there are.

Now, since when we consider the theory as a whole — a whole which includes the mathematics indispensable to the theory — we find mathematical entities among the things the theory says there are, we have that the mathematical theories which are indispensable to the successful theories belonging to the empirical sciences cannot be regarded as formal games.

To counter this objection the formalist can either try to show that mathematics is not indispensable to the empirical sciences or can reply that what explains the indispensability of a mathematical theory to a theory belonging to an empirical science is the interpretation provided of the language of the formal game/mathematical theory. 'In fact', the formalist would continue, 'it is such an interpretation what makes possible the setting up of a successful mathematical model of the phenomena studied by the theory'.

However, this reply is not very effective for two reasons. The first is that, as in the case of Field's programme for a nominalist reconstruction of physics, the attempts so far made to show that mathematics is not indispensable to physics have been unsuccessful.

The second is that, if we study the history of mathematics, we realize that the distinction operated by the formalist between a formal mathematical game and its interpretation is mistaken. For, such a distinction is the result of the attempt to transform the legitimate logical distinction between the two components of the language of a mathematical theory T — the syntax and the semantics of the language of T — into the unwarranted belief that T is a 'formal game' which coincides with the 'logical syntax of the language of T', and that the interpretation of the language of T is simply a device external to T which makes possible the application of T either to other formal games or to a theory belonging to the empirical sciences.

Indeed, the history of mathematics appears to reveal the existence of no cases of uninterpreted formal systems which have been accepted as legitimate mathematical theories by the mathematical community. On the contrary, in it may be found several instances of important mathematical theories which were originally developed to deal with the problems of some physical theory. One of the best known examples of this kind is the birth of the calculus at the hands of Newton.

4 Nominalism in mathematics

A contemporary nominalist is an empiricist philosopher who does not believe in the existence of abstract objects and who, in particular, does not believe in the existence of numbers, triangles, squares, sets, etc.

One of the main motivations of nominalism in mathematics is a generalization to all mathematical objects of the attitude Russell once had towards classes. For the Russell of *Introduction to Mathematical Philosophy* (**IMP**), in fact, classes were to be considered as mere fictions, because they are not 'part of the ultimate furniture of the world'. What Russell meant by this is explained in what follows:[9]

> If we had a complete symbolic language, with a definition for everything definable, and an undefined symbol for everything indefinable, the undefined symbols in this language would represent symbolically what I mean by "the ultimate furniture of the world." I am maintaining that no symbols either for "class" in general or for particular classes would be included in this apparatus of undefined symbols.

But how can the nominalist about mathematics justify his generalization of Russell's **IMP** view to all mathematical objects? Well, there are two main strategies followed by the nominalist to this end.

The first consists in showing that the view of mathematics which the nominalist considers to be contradictory with his own — Platonism — is untenable, because we can have no knowledge of abstract objects.

The second is both an attempt to undermine one of the main points in favour of Platonism — the Indispensability Thesis — and a way of providing a head on proof of the generalization of Russell's **IMP** view to all mathematical objects by showing, through a 'nominalistic reconstrual of mathematically formulated science and scientifically applicable mathematics',[10] that science can be done 'without numbers', sets, and any other abstract mathematical object.

The first strategy is realized by an argument which takes the following shape: since (P_1) whether a subject's true belief is knowledge depends on what caused the subject to have the belief, and (P_2) we can have no causal connections with abstract objects, it follows that (C) we can have no knowledge of abstract objects.

The argument above is clearly unsound, because premiss (P_1) is false, and the truth of premiss (P_2) is matter of dispute. Premiss (P_1) is false, because it conflates the context of discovery with that of justification. To see this, consider that what caused Newton to believe in universal gravitation was — so the story goes — the falling of an apple on his head. But, his true

---

[9][Russell, 1993], ch. XVII, p. 182.
[10][Burgess & Rosen, 1997], ch. I.A.o.a., pp. 5-6.

belief in universal gravitation became knowledge only when his system of mechanics provided logically compelling *reasons* to believe it true.

Secondly, premiss (P_2) is matter of dispute, because, although the Platonist deems it true, this is not so for the intuitionist. According to the intuitionist, abstract mathematical entities exist, but not independently of human activity. They are the outcome of those creative acts of the mathematician he calls 'mathematical constructions'.

Furthermore, a mathematical construction itself is an abstract entity. It is not above or below another construction, we cannot perceive it with our senses, but we can only understand/grasp it with our intellect. The abstract nature of mathematical constructions is particularly clear in the case of those mathematical constructions which can be carried out only in principle.

With regard to the second justificatory strategy of the nominalist, we have already come across the best example produced so far of its realization: Field's programme for the nominalistic reconstrual of physics.[11] And, in the course of the discussion of Field's programme, we realized the presence of faults within it which make impossible its realization.

However, even if it were possible to produce a 'nominalistic reconstrual of mathematically formulated science and scientifically applicable mathematics', this would not be a vindication of the nominalist's claim that there are no mathematical objects. For, since each nominalistic reconstrual of a scientific theory is a nominalistic reconstrual of a particular stage of development of the theory, a 'nominalistic reconstrual' of a particular stage S of development of a given scientific theory T would only show that in developing T up to S, we do not need to postulate abstract mathematical entities. Such a nominalistic reconstrual would, therefore, neither provide evidence in favour of the idea that we are not going to need to postulate the existence of abstract mathematical objects to bring T to stage $S + 1$, nor that there are no such entities.

From the arguments above follows that the only outcome of a positive deployment of the second nominalist strategy to a given stage of development of a scientific theory would bring about nothing more than a momentary weakening of the applicability of the Indispensability Thesis to that theory.

5 Fictionalism in mathematics

As it often happens in philosophy, several and contrasting are the standpoints associated by different authors with the same label. And, it is important to notice that the label 'fictionalism' is no exception to this rule. For

---

[11] An interesting description and discussion of strategies for the nominalistic reconstrual of physics alternative to Field's may be found in [Burgess & Rosen, 1997].

example, according to J. Burgess, mathematical fictionalism is that variety of nominalism[12]

> ... which helps itself to the utility of mathematics, while refusing to pay the price either of acknowledging that what mathematics appears to say is true, or of providing any reconstrual or reconstruction that would make it true.

On the other hand, for S. Hoffman, an acceptable fictionalist view of mathematics is one according to which mathematics is understood[13]

> ... as a set of stories about a fictional character [Kitcher's ideal mathematical agent] who resembles us in some ways and acts in a world somewhat like ours ...

In what follows I will be concerned with what in recent years has proved to be the most influential kind of fictionalist view of mathematics, that defended by H. Field.

For Field, a fictionalist in mathematics is someone who does not believe that mathematical statements are true, if they are taken at face value, even though, he[14]

> ... *may* believe that there is some non-face-value construal of mathematical sentences under which they come out true.

Therefore, when the mathematical fictionalist asserts 'Fermat's Last Theorem is true', what he means is 'According to standard mathematics, Fermat's Last Theorem is true' or 'Standard mathematics says that, i.e., has as a consequence that, Fermat's Last Theorem is true'.

Field's mathematical fictionalism is clearly an anti-realist philosophy of mathematics, because a fictionalist has no commitment whatsoever to the existence of numbers, sets, functions, etc.

An obvious objection to mathematical fictionalism is that such a view of mathematics cannot be right, because mathematical theories are indispensable for the development of the empirical sciences, etc. whereas this is certainly not the case with genuine fiction stories, even if some of them might have caught the imagination of more than one scientist.

To this objection, based on the so-called 'Indispensability Thesis', Field has a strong reply (see §2.10) which he has expounded at length in *Science without numbers*, and in several other contributions, a reply which aims at showing that, in contrast with the objection above, it is in principle possible to nominalize physical theories, i.e., to reformulate physical theories in such a way that no appeal to numbers, sets, functions, etc. is made within them.

---

[12][Burgess, 2004], §1, p. 18.
[13][Hoffman, 2004], §1, p. 4.
[14]See [Field, 1989], Introduction, p. 2.

However, we must realize that: (1) since Field's attack on the Indispensability Thesis is unsuccessful (see §2.11), the objection above against mathematical fictionalism stands; and that (2), even if Field were successful in carrying out his nominalization programme for physical theories, he would, as we noted in the case of the second justificatory strategy of the nominalist (§4.4), only produce a momentary weakening of the applicability of the Indispensability Thesis to physical theories. He would not succeed in showing that mathematical theories are not indispensable for physical theories.

6   Field's anti-objectivism

Field's view of mathematics has changed over the years. If his early contributions to the philosophy of mathematics focussed on whether or not mathematical objects exist, his most recent work can be said to revolve around the problem of mathematical objectivity.

The evolution of his thinking about mathematics has, probably, been the consequence of becoming convicted by the ideas expressed in the dictum that Dummett attributes to Kreisel, a dictum according to which, within the philosophy of mathematics, 'the problem is not the existence of mathematical objects, but the objectivity of mathematical statements.'[15]

The main aim of Field's later work in the philosophy of mathematics is developing a tenable anti-objectivist philosophy of mathematics.

According to Field, [16]

> ...the most objectionable feature of traditional platonism isn't the assumption of mathematical objects but its assumption that mathematics has a certain kind of objectivity. The kind of objectivity that I argue to be objectionable is the view that any sentence of mathematics has a determinate truth-value even if it is undecidable by current axioms or by any other axioms we are disposed to accept. The 'nominalism' or 'fictionalism' in my earlier work is simply one form that a properly anti-objectivist philosophy of mathematics can take.

In other words, a philosophy of mathematics is anti-objectivist, if it rejects the principle of Bivalence (see §1.7).

An interesting feature of Field's anti-objectivist position which distinguishes it, for example, from that of Michael Dummett is that, according to Field, one can be an anti-objectivist about mathematics without having to relinquish the use of classical logic in mathematical reasoning in favour of a non-classical one:[17]

> ...when I say that certain mathematical sentences might lack determinate truth value, I do not intend to suggest that we must abandon classical reasoning in connection with those sentences. In my view a great many concepts involve some sort

---
[15][Dummett, 1978], Preface, p. xxviii.
[16][Field, 2001], Preface, pp. ix-x.
[17][Field, 2001], Ch. 12, §2, p. 337.

of indeterminacy — for instance, vagueness — and as a result many sentences containing them lack determinate truth value. It would cripple our ability to reason if we were prevented from using classical logic whenever indeterminacy might arise. Fortunately, it is not necessary to do so: we can perfectly well say that everyone is either bald or not bald, as long as we add that not everyone is either determinately bald or determinately not bald. What's crucial to the logic of vagueness isn't that we give up classical logic but that we add to it a new 'determinately' operator — in effect, a notion of a sentence being determinately true. The same holds in the case of other sorts of indeterminacy. Consequently, standard mathematical reasoning can go unchanged when indeterminacy in mathematics is recognized: all that is changed is philosophical commentaries on mathematics, commentaries such as 'Either the continuum hypothesis is determinately true or its negation is determinately true.'

Now, in the debate surrounding the objectivity issue in mathematics, it is possible to distinguish between three different positions which emerge from the three possible answers that can be given to the question: which undecidable mathematical statements have determinate truth values?

The first position, that of the objectivist, answers 'all' to the question above; the second, that of the radical anti-objectivist, answers 'none'; and the third position, Field's, answers 'some'. In fact, Field believes that undecidable statements belonging to number theory are determinately true or false, but that certain undecidable statements belonging to set theory are not determinately true (false).

A moment's reflection on these issues makes us realize that Field's objectivism/anti-objectivism debate in mathematics is simply a way of presenting, in a slightly different terminology, the Dummettian realism/anti-realism debate about truth.

Indeed, for Field, the mathematical objectivist — like the Dummettian realist about truth — believes that all mathematical statements have a determinate truth-value; whereas the anti-objectivist — like a Dummettian anti-realist about truth — believes that there are some mathematical statements which do not have a determinate truth-value.

Therefore, regardless the differences existing between Field's and Dummett's ideas on: (1) the way their respective debates ought to be conducted; (2) whether classical logic ought to be expunged from mathematical reasoning; and (3) whether number theoretical undecidable statements have determinate truth values; it is correct to assert that if, as I argued in §§1.7–1.9, the issue concerning metaphysical realism is independent of that of realism about truth, it follows that, whichever of the three above mentioned positions on the objectivity issue is correct, this is not going to produce any information about the metaphysical issue of realism.

7 Constructivism: the mathematical aspect of verificationism

Verificationism is that form of empiricism which, originating from Kant's dictum that concepts without objects are empty,[18] believes in a dogma according to which the meaning of a statement consists in its method of verification.

It is interesting to notice that an immediate consequence of this dogma is that there are no absolutely undecidable statements. And, the reason why this is interesting is that the belief that there are no absolutely undecidable statements is philosophically very important, because it leads one to believe that those statements which can neither be known to be true nor be known to be false have, in fact, no meaning and should, therefore, be considered as pseudo-statements.

As is well known, verificationism became, at the hands of Carnap,[19] and Schlick[20] a very important part of the official doctrine of the Vienna Circle, and was used by Carnap and others as the philosophical basis for their systematic demolition campaign directed against metaphysics. The main weapon used in this campaign was a new one in philosophy: the logical analysis of language.

At this point some readers might start to wonder about what correlation there might be between constructivism and verificationism. The beginning of an answer to this question is that, for a constructivist, the truth of a mathematical statement does not depend on reality but on the existence of a constructive proof, that is, on the existence of a procedure of verification.

But, to see more clearly the connection existing between the idea of a constructive proof and that of verification, consider that constructivism, indeed, begins, as Bridges says, with the introduction of the view, common to some of the main constructivist mathematicians — Kronecker, Brouwer, Markov, and Bishop — that[21]

> the phrase *there exists* be interpreted strictly as *we can construct*

and that the phrase *we can construct* should be understood as *we can compute*.

From what has just been said, we obtain a clear idea of what a constructive proof of an existential statement consists in. For instance, a constructive proof of the statement

[a] There exists a natural number n such that $P(n)$

---

[18] See on this the quotation given in §3.4, p. 91.
[19] See [Carnap, 1932].
[20] See [Schlick, 1930].
[21] [Bridges, 1998], §1, p. 53.

consists in producing an algorithm $\mathcal{A}$ which enables us to compute a natural number c and show that $P(c)$.

The above mentioned schema of a proof procedure of [a] is a verification procedure for [a], because the eventual proof actually finds a natural number c and shows that c falls under the concept P. Not so with regard to the proof procedure which, starting from the assumption ¬[a], derives a contradiction.

From a generalization of these ideas to a constructivist understanding of the logical constants, for how this is given by the BHK-interpretation, we can gain a clear picture both of the connection existing between constructivism and verificationism, and of the reason why the constructivist finds objectionable in the reasoning of classical mathematics (CLASS) what has become known as 'giving proofs by logic'.[22]

In attempting to give a more detailed characterization of constructivism, it is important to notice that in the equivalence established by the constructivist between existence and computability, the very concept of computability must not be interpreted according to classical mathematics.

Let us see how Bridges explains this point:[23]

> Consider the following example of a function f from the set $\mathbb{N}$ of natural numbers to itself:
> $$f(n) = \begin{cases} 0 & \text{if the Goldbach conjecture is false,} \\ 1 & \text{if the Goldbach conjecture is true.} \end{cases}$$
> (The Goldbach Conjecture states that every even integer greater than 2 is a sum of two primes.) In classical mathematics this is regarded as a computable function,

---

[22] As a matter of historico-philosophical interest, it is worth pointing out that it was the verificationism to be found at the very roots of this view of mathematics what led the constructivists to consider unsound the classical interpretation of the Axiom of Choice, even before it became known that the Axiom of Choice implies the law of excluded middle. See [Goodman & Myhill, 1978]. In what follows I will give a quotation from [Bridges & Richman, 1988], Ch. 1, §4, pp. 11-12, where the argument to the effect that the Axiom of Choice implies the Law of Excluded Middle is expounded:

> The full axiom of choice reads as follows.
>
> **Axiom 7.1.** If S is a subset of $A \times B$, and for each x in A there exists y in B such that $(x, y) \in S$, then there is a function f from A to B such that $(x, f(x)) \in S$ for each x in A.
>
> The function f is called a **choice function** for S. Goodman and Myhill have shown that this axiom implies the law of excluded middle ... Their argument establishing $P \lor \neg P$ is very simple. Let $A = \{s, t\}$, where $s = t$ if and only if P; let $B = \{0, 1\}$; and let $S = \{(s, 0), (t, 1)\} \subset A \times B$. If $f : A \to B$ is a choice function for S, then either
>
> (i) $f(s) = 1$ or $f(t) = 0$, so that $s = t$, and therefore P holds; or else
>
> (ii) $f(s) = 0$ and $f(t) = 1$, so that s cannot equal t, and therefore P cannot hold □

[23] [Bridges, 1998], pp. 53-54.

7. CONSTRUCTIVISM 141

since there exists (in the idealistic sense — it is absurd that there not exist) an algorithm that, applied to any natural number n, outputs $f(n)$. In fact, there are two algorithms — one, $\mathcal{A}_0$, that always outputs 0, and one, $\mathcal{A}_1$, that always outputs 1 — one of which must, classically, compute f.

Now, I find it strange to describe a function $f : \mathbb{N} \to \mathbb{N}$ as computable when, with our present state of mathematical knowledge, we cannot even compute its value at the input 0 ...

Is it, then, sensible to call such functions as $f \ldots$ 'computable'? The constructivist would say

No. It only makes sense to call f computable if we can decide which of the two algorithms $\mathcal{A}_0$ and $\mathcal{A}_1$ computes f ... In other words, we are only justified in calling f computable if we can decide the Goldbach conjecture.

Another important characteristic of constructivist mathematics is its criticism of the use made within CLASS of infinite sets as completed totalities. Indeed, the constructivist argues, if existence claims are equivalent to claims about the possibility of effecting computations, it follows that we are justified in speaking of the existence of a set A just in case we can produce a computable function f whose range is A.

Therefore, we can talk of a set A in terms of a *completed totality* only if the computation of the values of f comes to an end. But, if this is not the case, in the sense that there is no stopping the growth of the range of f, then our computable function f would only generate a potentially infinite collection. From this follows that, for the constructivist, potential infinity is the only meaningful concept of infinity to be used in mathematics.

It is important to notice that the acceptance, within constructivism, of the concept of potential infinity and the consequent rejection of that of actual infinity determine, more than the differences existing between constructivist and classical logic,[24] a profound conflict between constructivism and classical mathematics, as it is shown by the fact that one of the monuments of classical mathematics, the theory of transfinite cardinal numbers, is not constructively acceptable.

Having proceeded to a preliminary discussion of some of the tenets of constructivism, it should be clear by now that constructivism is an anti-realist philosophy of mathematics according to which mathematics is a creative activity of the human mind, i.e., mathematics consists in the activity of creating/inventing mathematical constructions.

However, constructivism is a broad church within which we find several schools of thought which differ from one another in relation to how radical their philosophical ideas and/or programmes for a reformation of mathematical practice are.

---

[24]Constructive logic coincides with intuitionistic logic, and there is a result (the negative translation of classical logic into intuitionistic logic) which shows that classical and intuitionistic logic are, in fact, equi-consistent.

Indeed, constructivism, in contrast with other anti-realist views of mathematics like formalism and Field's anti-objectivism, is characterized by a strong revisionary zeal which aims at reforming classical mathematics by purging it from all the mythological accretions which have affected, among other things, its view of mathematical proof and truth.

In what follows I shall examine some of the most important schools of constructivist thinking in mathematics.

8 Brouwer's Intuitionism

One of the oldest and most influential brands of constructivism is Brouwer's Intuitionism. Brouwer's Intuitionism takes its inspiration from Kant's philosophy of mathematics, an inspiration which leads Brouwer to develop not just another Kantian philosophy of mathematics, but a research programme concerning the foundations and the practice of mathematics which challenges classical mathematics.

For Brouwer, the activity of producing mathematical constructions is a creative activity of the mind. According to him, language plays no part in the process leading up to the invention of constructive proofs (or refutations), its only function being that of recording mathematical constructions and communicating them to other people.

But, if mathematical constructions are not to be conceived as subjective processes of the mind, which magically assume a public, objective form becoming expressible in language and, therefore, communicable to other people, they must be based on some faculty of reason which makes them intersubjective. If this is the case, it is legitimate to ask which faculty of reason this is, and whether the presupposition of this faculty of reason is sufficient to save the objectivity of mathematics. To answer these questions we need to take a step back and engage in a more systematic discussion of Brouwer's Intuitionism.

The starting point of Brouwer's reflections on mathematics is the problem of explaining the unassailable exactness of mathematical laws. And, with regard to this problem, he says that mathematical exactness exists in the intellect adopting a Kantian position in which, as a consequence of the existence of non-Euclidean geometries, the *a priori* nature of space must be abandoned, but that of time upheld.

Brouwer's Intuitionism[25]

> ...considers the falling apart of moments of life into qualitatively different parts, to be reunited only while remaining separated by time, as the fundamental phenomenon of the human intellect, passing by abstracting from its emotional content into the fundamental phenomenon of mathematical thinking, the intuition of the

---

[25][Brouwer, 1912], p. 80.

8. BROUWER'S INTUITIONISM

bare two-oneness. This intuition of two-oneness, the basal intuition of mathematics, creates not only the numbers one and two, but also all finite ordinal numbers, inasmuch as one of the elements of the two-oneness may be thought as a new two-oneness, which process may be repeated indefinitely; this gives rise still further to the smallest infinite ordinal number ω. Finally this basal intuition of mathematics, in which the connected and the separate, the continuous and the discrete are united, gives rise immediately to the intuition of the linear continuum, i.e., of the "between", which is not exhaustible by the interposition of new units and which therefore can never be thought of as a mere collection of units.

If we consider with some attention the quotation above, we will find important elements in it which clash with Kant's own view of number. The most striking of these is, in my view, what Brouwer calls 'the intuition of the bare two-oneness', intuition which appears to be the faculty of reason that, for Brouwer, provides mathematical constructions with an intersubjective foundation.

The first thing we must notice is that, according to Brouwer, the intuition of the bare two-oneness is the outcome of a process of abstraction from our temporal representations of moments of life, a process of abstraction from which are generated all finite ordinal numbers; whereas, for Kant, the natural numbers are not generated by a process of abstraction at all.

Indeed, according to Kant, the natural numbers are neither created by intuition nor by anything else, they are pure *a priori* schemata of magnitude which *presuppose* the categories of quantity and make possible the application of such categories to the imagination. What this means is that, when we think of a natural number n, however large this may be, we do not have an image in mind, but an *a priori* method — adding one entity to a set of entities homogeneous with it — whereby n may be represented in an image of n dots or n exclamation marks, etc.

Therefore whereas, for Kant, mathematics is a genuine science, whose judgments are synthetic *a priori*, a science which produces information concerning the existence and properties of numbers, geometrical figures, etc. according to Brouwer, mathematics is a purely creative activity of the subject.

Secondly, Brouwer's Intuitionism is a variety of mathematical constructivism and, as we have seen in the previous section, mathematical constructivism is a version of verificationism applicable to mathematics. This is important for our discussion, because verificationism in general, and mathematical constructivism in particular, make sure there are no meaningful (mathematical) concepts without objects.

Such a feature of mathematical constructivism, inherited by Brouwer's Intuitionism, is very useful in understanding some features of Brouwer's contributions to the philosophy of mathematics. For, it explains both why Brouwer felt he was providing mathematical flesh and blood to Kant's philosophy of mathematics; and why he thought mathematical conjectures to be devoid of sense. But, let us now consider some more specific features of Brouwer's Intuitionism.

According to Brouwer, saying '$P(x)$ is true' means 'There is a constructive proof of $P(x)$', and, as Heyting says:[26]

> **Int**: ... Every mathematical assertion can be expressed in the form: 'I have effected the construction A in my mind'. The mathematical negation of this assertion can be expressed as 'I have effected in my mind a construction B, which deduces a contradiction from the supposition that the construction A were brought to an end', which is again of the same form.

But what do we have to understand by 'constructive proof' or 'construction', in Brouwer's sense? Is this a notion which can be entirely captured by what I said on this topic in the previous section?

Well, the answer to these questions cannot be entirely straightforward. In fact, although Heyting produced an interpretation of the logical constants, which is currently accepted as a fair embodyment of the intuitionist's ideas about first-order logic (this is the BHK-interpretation), the Brouwerian notion of mathematical construction, as involving vastly more than logic, has been taken as primitive by the faithful followers of Brouwer. The reason for this being that, since constructions are creative acts of the subject, we can distinguish between them, as van Stigt says, 'genetically', but we cannot provide a complete classification of constructions. This situation is determined not so much by the fact that *there are too many* of them (constructions) or by the inadequacy of ways of formalising constructions, but by the *openness* of the notion of construction which is made to depend on the vague notion of 'creative activity' of the subject.[27]

As is well known, in spite of all the methodological differences, if we stay with finite sets, intuitionists and followers of Hilbert's programme are in complete agreement. As Brouwer himself says:[28]

---

[26][Heyting, 1966], §2.2.2, p. 19.
[27][van Stigt, 1990], Ch. IV, §4.5.2, p. 167:

> The lack of simplicity and uniformity in the domain of mathematical constructions is mainly due to the freedom of the Subject to create ever more and more complex constructions, using Intuition and previously constructed entities and tools in 'a free unfolding'. The condition 'previously constructed' is not just a restriction, it is a sanction of their mathematical birth-right to be exploited to the full. For example, [Brouwer, 1912], points out that in the construction of the infinite ordinal ω 'of course, every previously constructed set of every previously performed constructive operation may serve as the unit'. In this free-unfolding, however, the 'previously acquired' can only be a general principle for a genetic hierarchy of complex constructions, and was used as such e.g. in Brouwer's hierarchy of species (see 6.3.7). A comprehensive hierarchical classification of all mathematical constructions must remain illusory.

[28][Brouwer, 1912], p. 82.

8. BROUWER'S INTUITIONISM

> In the domain of finite sets in which the formalist axioms have an interpretation perfectly clear to the intuitionists, unreservedly agreed to by them, the two tendencies differ solely in their method, not in their results; this becomes quite different however in the domain of infinite or transfinite sets, where, mainly by the application of the axiom of inclusion, quoted above, the formalist introduces various concepts, entirely meaningless to the intuitionist, such as for instance '*the set whose elements are the points of space,*' '*the set whose elements are the continuous functions of a variable,*' '*the set whose elements are the discontinuous functions of a variable,*' and so forth.

However, the difference between Brouwer's approach and that of Hilbert's finitary mathematics (see §4.3) emerges unexpectedly already in number theory. In fact, if we consider the result proved in [Gödel, 1933e] concerning the consistency of a first-order formal system of classical arithmetic (**PA**) relative to a first-order formal system of intuitionistic arithmetic (**HA**), we realize that what Gödel has produced is an intuitionistically acceptable proof of the consistency of **PA**. From this result and Gödel's second incompleteness theorem, we must conclude that the proof of the consistency of **PA** offered by Gödel in [Gödel, 1933e] cannot be finitistic in Hilbert's sense and that, consequently, some of the intuitionistically acceptable methods used in it go beyond Hilbert's finitary mathematics.

But, of course, since Gödel's above mentioned result can be used to show that **HA** and **PA** are equi-consistent, all those who do not believe in the self-evident consistency of **HA** will say that what has been proved by Gödel in [Gödel, 1933e] is a confirmation of their belief that **HA**, as well as **PA**, is in need of a consistency proof.

Commenting further on intuitionistically acceptable methods, it is important to mention that Brouwer considers legitimate those mathematical constructions which can be carried out *in principle*, i.e., independently of considerations concerning resources; and that he allows mathematical constructions which are not determinate in advance by a *law*.

With regard to the difference between intuitionistic and classical mathematics, it is important to say that inuitionistic analysis, far from being similar in the meagreness of the outcome to the activity of 'a boxer deprived of the use of his fists', has developed, at the hands of Brouwer and others, into a rich and interesting theory. It adds to the interest of intuitionistic analysis that this is not a sub-theory of classical analysis in that some of its theorems, like the Uniform Continuity Theorem I mentioned in §1.8[29] are classically false.

After the brief discussion offered above of some of its main ideas, time has now come to ask whether Brouwer's Intuitionism succeeds in solving

---

[29]The interested reader may consult on this standard texts such as [Dummett, 2000], Ch. 3, §3.6, pp. 81-87.

the problem of explaining the unassailable exactness of mathematical laws, and in expressing a viable philosophy of mathematics.

Concerning the problem of establishing the unassailable exactness of mathematical laws, Brouwer's Intuitionism seems to lead to the grotesque and unwanted conclusion that many mathematical laws are *not* unassailable after all. In fact, how else should we interpret the intuitionist's assault on and rejection of the constructivist equivalent of the law of excluded middle and of many other classically valid statements and techniques of proof, including a large part of CLASS-analysis?

To this objection it is possible to reply that there is no real conflict between Brouwer's main aim in the philosophy of mathematics and the presence within Intuitionism of a reforming zeal whose objective is not simply to prune CLASS from any gratuitous flight of fancy, but to put mathematics on a completely different track. For, what Brouwer is really arguing in favour of is the unassailable exactness of the laws which belong to *intuitionistic mathematics*.

However, this counter-objection turns out not to be compelling, because the exactness of the laws which belong to intuitionistic mathematics has been challenged not only, as one would expect, by classical mathematicians, but also by various constructivist philosophers of mathematics.

One of the main objections against the exactness of the laws and methods of intuitionistic mathematics moved from within the constructivist's camp is that it is constructively unacceptable to consider as legitimate mathematical constructions which can be carried out in principle, but which, as a consequence of the lack of resources, cannot be actually carried out in most cases.[30]

Indeed, since Intuitionism is a variety of verificationism applied to mathematics, for the constructivist critics of Intuitionism, the intuitionist's acceptance of a decision procedure which cannot be actually carried out in most cases has to be regarded in the worst scenario as a contradiction in terms and in the best as just depending on the myth of a resources free creative subject able to carry out computations of indefinitely large complexity, a myth which clearly betrays the verificationist philosophy on which Intuitionism stands.

This constructivist complaint against intuitionist mathematics which, of course, extends to the mathematical meaningfulness of processes which 'may

---

[30] An example of this phenomenon is the usual algorithm for determining whether a natural number n is prime or composite. In fact, for an n sufficiently large, there would not be enough resources in the universe to implement the algorithm. And, of course, being $I_n = \{x \mid x \in \mathbb{N} \text{ and } x < n\}$ a finite set, it follows that an application of the algorithm to most natural numbers would not actually yield an answer to the question whether they are prime or not.

8. BROUWER'S INTUITIONISM

be repeated indefinitely' and 'give rise still further to the smallest infinite ordinal number ω', resonates with the spirit of Brouwer's remarks on the meaninglessness (to the intuitionist) of the introduction of concepts such as 'the set whose elements are the points of space', etc. In the case of Intuitionism, the object of the constructivist's criticism is, of course, the mathematical meaningfulness of indefinitely long constructions such as that which generates ω.

Another point that needs attention is the following. Earlier in this section I said that, for Brouwer, asserting '$P(x)$ is true' means asserting 'There is a constructive proof of $P(x)$'. From this we can derive that, for Brouwer, the notion of truth can be reduced to that of constructive provability. The reduction of the concept of truth to that of constructive provability is vey important for the intuitionist, because, if things concerning mathematical truth were otherwise, the way would be wide open for a realist account of mathematical truth, that is, for an account of mathematical truth as verification transcendent.

But now a legitimate question arises: can we understand the notion of proof independently of the notion of truth? To answer this question we should ask ourselves when we say in mathematics that a proof Π of a mathematical statement S from a set of assumptions $A_1, \ldots, A_n$ is correct. Well, Π is correct, if it shows that whenever we assert $A_1 \wedge \cdots \wedge A_n$, we can assert S as well, otherwise Π is incorrect.

Therefore, independently of the methods used within Π, the concept of truth turns out to be an indispensable criterion to decide whether Π is a proof or not. For, as I argued above, what a proof does is revealing the existence of a relation of logical dependence of the proved statement S on the set of assumptions used in the proof of S. From this follows that the concept of truth in mathematics cannot be reduced to that of mathematical proof.

Brouwer's Intuitionism may also be criticized on the grounds that, since the mathematical activity of the creative subject rests on the psychological phenomenon of the temporal representation of 'moments of life', it follows that Brouwer's Intuitionism, being inextricably bound with psychologism, cannot save mathematical objectivity.

For, apart from the inherent vagueness present in a fundamental concept such as that of 'moments of life', if mathematics is a creative activity which entirely rests on a psychologistic basis, on *inner* phenomena, how can we justify believing that Brouwer's temporal representation of 'moments of life' is the same as that of any other mathematician, and that our own present temporal representation of a moment of life is the same as the temporal representation of that moment of life we remember we had in the past?

(Wittgenstein's private language argument.)

Lastly, it seems unreasonable to assert, as Brouwer does, that mathematical conjectures are meaningless. Indeed, although at first sight it seems plausible for a very strict verificationist like Brouwer to believe that mathematical statements lacking a criterion of verification, that is, a mathematical proof or a refutation, are meaningless, on the other hand, it is impossible to understand how, in particular for an anti-realist philosopher of mathematics, the proof (refutation) of a mathematical statement S might not depend on the meaning of S, and how a proof (or a refutation) of S offered at time t can transform a meaningless phrase into a meaningful mathematical statement.

9 Dummett's Intuitionism

Dummett's Intuitionism comes about as the attempt to eliminate two of the most serious defects present in Brouwer's Intuitionism. The first of these defects is the psychologism that pervades Brouwer's account of mathematics and paves the way for the slide towards mathematical subjectivism and solipsism. The second is Brouwer's attitude towards mathematical conjectures.

The main point of departure of Dummett's Intuitionism from Brouwer's is to be found in the rôle Dummett assigns to language in mathematics. Indeed, Dummett's quasi-Wittgensteinian view of language, for which meaning is neither a private mental entity nor an abstract object, but must be accounted for in terms of use, considers language in general, and the language of mathematics in particular, to be public entities based on convention and regulated by public criteria of correctness.

Now, if mathematics is an activity which can take place only in language, it becomes clear how Dummett can avoid the intrusion of psychologism in intuitionistic mathematics. In fact, since language is a public, objective entity, it follows that mathematical constructions must also be both public and objective.

Furthermore, Dummett's *linguistic turn* operated within Intuitionism, taken together with a molecularist view of the language of mathematics, succeeds in disposing of Brouwer's implausible attitude towards mathematical conjectures.

To see this consider that, for Dummett, we can call 'molecular'[31]

> ...any view [of language] on which individual sentences carry a content which belongs to them in accordance with the way they are compounded out of their own constituents, independently of other sentences of the language not involving those constituents.

---

[31][Dummett, 1973], p. 104.

9. DUMMETT'S INTUITIONISM

Consequently, if you are a molecularist about, for example, the language of arithmetic then, for you, Goldbach's conjecture, the Riemann hypothesis, etc. have a content, a meaning 'in accordance with the way [Goldbach's conjecture, the Riemann hypothesis, etc.] are compounded out of their own constituents, independently of other sentences of the language not involving those constituents' and, in particular, independently of whether or not we have an effective procedure of decision concerning their truth.

One of the most interesting features of Dummett's linguistic turn is that, besides providing an efficacious defense of Intuitionism against two very strong objections, it bases the justification of Intuitionism and the rejection of classical logic and of Platonism in mathematics, on considerations which have to do with the meaning and the truth of mathematical statements rather than with speculations belonging to traditional metaphysics. But, let us now see how Dummett deploys his argument against classical logic.

Dummett's first move consists in justifying the acceptance of the Wittgensteinian principle that the meaning of a mathematical statement determines and is exhaustively determined by its use.

To do this he assumes as self-evident the principle according to which (S_1): the meaning of a mathematical statement consists only in its rôle as an instrument of communication between individuals; and then argues that (S_2): if an individual associates with a mathematical symbol or formula some mental content, and such an association does not lie in the use made of the symbol or formula, then he cannot convey that content by means of the symbol or formula and, consequently, his audience would have no means of becoming aware of it.

However, since (S_3): — hidden assumption — in ordinary communication we are aware of the content of mathematical statements, it follows that (C): the meaning of a mathematical statement determines and is exhaustively determined by its use.

At this point Dummett is ready to take on board (C) in the construction of his main argument. But, before giving a description of the argument, it is important to keep in mind that this is aimed at showing that an interpretation of the language of a mathematical theory based on truth-conditional semantics violates principle (C).

Dummett begins his main argument stating that (M_1): the Platonist is committed to accepting truth-conditional semantics, that is, a semantics based on the idea that grasping the meaning of a statement belonging to the language of a mathematical theory consists in knowing what it is for that statement to be true.

He then considers those statements belonging to a mathematical theory which are not effectively decidable ($e.d.$), that is, those statements for which we do not have an effective procedure of decision concerning their truth, and says that (M_2): the Platonist defender of truth-conditional semantics claims to know what it is for non-$e.d.$ statements to be true, in the absence of a manifestable ability (use) to recognize which of them are true.

Hence,[32] (C^*):

> ...the knowledge which is being ascribed [by the Platonist] to one who is said to understand the [non-$e.d.$] sentence is knowledge which transcends the capacity to manifest that knowledge by the way in which the sentence is used.

There are two things we must notice at this point. First, (C^*) implies that[33]

> The platonistic theory of meaning cannot be a theory in which meaning is fully determined by use.

Secondly, the one above is a conclusion which has as consequence that a truth-conditional meaning theory is not adequate to the language of mathematics.

From these considerations it follows that, if Dummett's main argument is valid, and we want to produce a meaning theory adequate to the language of mathematical theories, we must reject truth-conditional semantics and classical logic, in favour of a different type of semantics and of a non-classical system of logic.

This situation appears to have serious metaphysical consequences. For, if Dummett is right in saying that the Platonist is committed to truth-conditional semantics and classical logic, and truth-conditional semantics and classical logic are inadequate for the language of mathematics, Platonism turns out to be inadequate as well, as a philosophy of mathematics.

Important components of the *pars construens* of Dummett's discussion of the philosophical basis of classical and intuitionistic logic are: (a) a definition of what it is to understand the meaning of a mathematical statement which appeals to the concept of provability[34] rather than to that of truth;[35]

---

[32][Dummett, 1973], p. 107.
[33][Dummett, 1973], p. 107.
[34][Dummett, 1973], p. 107:

> ...consists in a capacity to recognize a proof of it when one is presented to us, and a grasp of the meaning of any expression smaller than a sentence must consist in a knowledge of the way in which its presence in a sentence contributes to determining what is to count as a proof of that sentence.

[35][Dummett, 1973], p. 129:

> ...there is no notion of truth applicable even to numerical equations save that in which a statement is true when we have actually performed a computation (or effected a proof) which justifies that statement.

9. DUMMETT'S INTUITIONISM   151

and (b) a case made in favour of the idea that verificationist meaning theories and intuitionistic logic succeed where truth-conditional meaning theories and classical logic fail.

Concerning the *pars destruens* of Dummett's discussion, we must notice that, if what I have argued in §§, 1.7–1.9 is correct, it follows that, since the issues about metaphysical realism and realism about truth are independent of one another then, in contrast with assumption (M_1), the Platonist about mathematics is committed neither to a truth-conditional semantics nor to classical logic.

Secondly, to the idea that 'implicit knowledge [of meaning] can be sensibly ascribed to a person, only if it is fully manifestable', we can object that this is a very dubious claim to make even when we consider effectively decidable statements and predicates.

To see this, take as an example the predicate (P) 'n is prime', where $n \in \mathbb{N}$. Now, producing a definition of prime number is, of course, not sufficient as a manifestation of knowledge of the meaning of (P), because the problem would at this point become 'Does he understand the meaning of the definition he has just given?' etc. launching an infinite regress.

Therefore, the only way of manifesting a knowledge of the meaning of (P) is in the act of recognizing certain numbers as satisfying (P) and others as not satisfying (P). But, since n varies over an infinite domain, the problem of determining whether someone has understood the meaning of (P) is only partially decidable by what this person does. For, if, in normal circumstances, he falsely asserts that n is prime, for some $n \in \mathbb{N}$, then we can say that he has not understood the meaning of (P), but if, on the other hand, he correctly asserts that (P), we are not in a position to tell whether he has understood (P) or not, because it is possible that, for a $k \in \mathbb{N}$ sufficiently large, he might, in normal circumstances, falsely assert that 'k is prime'. And the fact that in practice, after certain tests (exams), we settle (un)officially the question whether or not X understands the meaning of (P) means nothing more than 'What we have seen of X's performance has satisfied us concerning whether or not he has a grasp of the meaning of (P)'

Thirdly, as we have seen in §1.8, the result according to which, for the intuitionist, there are no absolutely undecidable mathematical statements puts the Dummettian intuitionist before the two horns of a dilemma — introducing tense in mathematics or becoming a realist about proofs — either of which is unacceptable to him.

Indeed, we there noticed that introducing tense in mathematics is unacceptable for the intuitionist, and that, on the other hand, a realism about proof *à la* Prawitz, is incoherent with the anti-realist *credo* of Intuitionism. Dummett sees very clearly this last point, as it emerges in the following quotation:[36]

> There is a well-known difficulty about thinking of mathematical proofs — and equally, of verifications of empirical statements — as existing independently of our hitting on them, which insisting that they are proofs we are capable of grasping or of giving fails to resolve. Namely, it is hard to see how the equation of the falsity of a statement (the truth of its negation) with the non-existence of a proof or verification can be resisted: but, then, it is equally hard to see how, on this conception of the existence of proofs, we can resist supposing that a proof of a given statement determinately either exists or fails to exist. We shall then have driven ourselves into a realist position, with a justification of bivalence.

Of course, to this remark of Dummett Prawitz can reply that, although the admission, on the intuitionist's part, that proofs exist before we find them justifies the classical reading of the principle of bivalence, the intuitionist can still reject *his* reading of the law of Excluded Middle, because, for an arbitrary statement S, we do not have a procedure which allows us to find either the proof of S or that of $\neg S$. But what good would this do to the Dummettian debate and to the verificationist faith of the intuitionist? What good would it do to the cause of Intuitionism in general to replace the Platonist's faith in the existence of a 'verification independent' mathematical reality populated by abstract objects such as numbers, geometrical figures, and abstract structures, with the faith in the existence of a 'verification independent' realm of proofs?

10 Markov's constructivism

Although Intuitionism is the most philosophically sophisticated form of constructivism, it is by no means the only one. As we shall see in what follows in this chapter, the nature and the extent of the disagreement existing between the different schools of constructive mathematics will reveal that constructivism is unable to save a fundamental feature of mathematics, its objectivity.

Let us begin our brief survey of the main non-intuitionist strands of constructive mathematics with a discussion of some of the ideas of Markov (RUSS). Markov's constructivist mathematics has at its heart the concept of effective (mechanical) computability, which comes together with a *de facto* acceptance of Church's Thesis (CT): the class of computable functions coincides with the class of recursive functions.

For Markov, the most basic objects to which effective processes apply are the natural numbers and the recursive functions. Both the natural numbers and the recursive functions are seen by him as well-formed strings of symbols (*words*) on an alphabet.

---

[36][Dummett, 1987], p. 285.

10. MARKOV'S CONSTRUCTIVISM

Indeed, if we adopt the decimal notation, the natural numbers can be seen as words on the alphabet $\mathcal{A} = \{0, 1, \ldots, 9\}$. (In more general terms, the natural numbers can be thought as words on an alphabet $\mathcal{A}$ containing at least one element.) With regard to the recursive functions, if these are expressed by a list of equations, they also appear to be objects of Markovian effective processes, because equations are words on a given alphabet.

According to Markov, any effective process applicable to words on an alphabet $\mathcal{A}$ can be obtained combining elementary effective processes on words belonging to the same alphabet. Such effective processes are specified by Markov's *normal algorithms*, which are finite sequences of rewriting (substitution or production) rules very much like those of a context-free grammar.[37]

Furthermore,[38]

> More complex objects are represented by Gödel numbers; so, for example, a function between sets of recursive reals is represented by a function between the corresponding sets of Gödel numbers.

Lastly, it is important to notice that the system of logic adopted by Markov is intuitionistic; and that one of the consequences of Markov's view of mathematics — which is false in CLASS + CT — is that the number-theoretical functions are recursive.

Having laid out some of the main ideas of Markov's constructivism, it is now important to focus on the differences between Markov's constructive mathematics and Intuitionism.

As Troelstra and van Dalen say,[39]

> On the one hand Markov is more strict than the intuitionists, e.g. he rejects choice objects, [and] on the other hand he is more liberal. This is illustrated by his argument for [the] justification of his *principle of constructive choices*, nowadays called *Markov's principle* ...

In contrast with what has been asserted in the quotation above, we must observe that Markov's rejection of free-choice sequences is not a true departure point from Intuitionism, because there are brands of Intuitionism which do not look with sympathy on free-choice sequences either.

A very different matter is, instead, represented by Markov's principle (**MP**). According to (**MP**), if H is an algorithmically decidable number-theoretical (unary) predicate, i.e., for any natural number n there exists an algorithm which establishes either $H(n)$ or $\neg H(n)$, then

$$(\mathbf{MP}) \qquad \neg\neg\exists x H(x) \to \exists x H(x).$$

---

[37]See on this: [Cutland, 1988], Ch. 3, §5; [Hermes, 1965], Ch. 7, §33.2, pp. 233-234; [Mendelson, 1987], Ch. five, §5.II, pp. 268-271.

[38][Bridges, 1998], §2, p. 55.

[39][Troelstra & van Dalen, 1988], **vol. I**, Ch. 1, §. 4.6, p. 27.

MP, which is, of course, anathema for the intuitionist, has a very plausible justification, for Markov, given in the following terms. If we have a proof that it is impossible that there is no $n \in \mathbb{N}$ such that $H(n)$, apply the following algorithm: (i) take 0 and apply the H-algorithm to it. If $H(0)$ then stop; if $\neg H(0)$, (ii) go to 1 and repeat the procedure above, i.e., apply the H-algorithm to 1, etc. Since $\mathbb{N}$ is denumerable and well-ordered under the natural ordering, after a finite number of steps, the algorithm just described will hit against an $n \in \mathbb{N}$ and show that $H(n)$, hence, giving a constructive proof of $\exists x H(x)$.

Markov's principle can also be formulated in terms of Turing machines as stating[40]

> ...that if it is not the case that a Turing machine does not halt on a given input, then it does halt. Markov's argument, in terms of Turing machines, is simply as follows; if it is impossible that the Turing machine will compute for ever, then there is a clear algorithm for obtaining the output: just continue the process until it halts.

Troelstra and van Dalen give a very illuminating quotation from Markov in which he justifies the introduction of his principle. This quotation is particularly interesting for us, because it points clearly out, both with its tone and its content, *the* insuperable obstacle that constructivism has in elaborating a tenable : the lack of objectivity. But let us see what Markov says:[41]

> I know that intuitionists reject this method of argument, since they do not consider it 'intuitively clear'. In connection with this I consider it necessary to make the following remarks. Firstly, my intuition finds this sufficiently clear. On the other hand, I can in no way agree to taking 'intuitively clear' as a criterion for truth in mathematics, for this criterion would mean the complete triumph of subjectivism and would lead to a break in the understanding of science as a form of social activity. If I defend this means of argument here, it is not because I find it without error according to my intuition, but rather, firstly because I see no reasonable basis for rejecting it, and secondly, because arguments of this type make it possible to construct a constructive mathematics that is well able to serve contemporary natural science.

Before moving on to the study of another variety of constructivist philosophy of mathematics, I must mention that RUSS analysis is incompatible with both CLASS and INT analysis. An example of this is given by the following RUSS theorem:[42]

Theorem 10.1. There exists a pointwise continuous function from $[0, 1]$ onto $(0, 1)$ that is not uniformly continuous.

---

[40][Troelstra & van Dalen, 1988], ibid.
[41][Markov, 1962] in [Troelstra & van Dalen, 1988], ibid.
[42][Bridges & Richman, 1988], Ch. 3, §3, p. 59.

11. BISHOP'S PHILOSOPHY OF MATHEMATICS 155

Now, theorem **10.1** clearly contradicts the Uniform Continuity Theorem of INT, but also contradicts the following CLASS theorem:

Theorem 10.2. If f is a continuous function from a compact metric space (X, d) into a metric space (Y, d^*), then f is uniformly continuous.

In fact, since the interval of reals $[0, 1]$ is compact (Heine-Borel theorem), and is a metric space with metric $d(a, b) = |a - b|$; and the interval of reals $(0, 1)$ is a metric space with metric $d(a, b) = |a - b|$, it follows from theorem **10.2** that *any* pointwise continuous function

$$f : ([0,1], d) \to ((0,1), d)$$

is uniformly continuous.

From what we have seen so far of RUSS and INT, we can conclude that RUSS and INT are two types of constructivism which are incompatible with one another. However, it seems to me that, over and above the important logico-mathematical differences, there is a profound philosophical gulf separating RUSS from INT.

In fact, if, on the one hand, some fundamental features of RUSS such as the emphasis placed on effective procedures, the *de facto* acceptance of Church's thesis, and the types of objects to which effective procedures are applicable, appear to force upon us a picture of mathematical activity which is very much like that of a machine; on the other hand, for the intuitionist, the concept of mathematical construction is open ended, so much so that Brouwer accepts non-lawlike objects such as free-choice sequences, and constructions based on them.

Therefore, when we consider INT and RUSS, we are not simply facing two different, mutually incompatible constructive frameworks for mathematics. When we consider INT and RUSS, we are also before two different, mutually incompatible, verificationist philosophies of mathematics.

11 Bishop's philosophy of mathematics

Another variety of constructivism is represented by the work of E. Bishop and his followers. For Bishop, mathematics is a creation of the human mind which has as its primary concern '...number, and this means the positive integers'.[43] The reason for numbers being, according to Bishop, the primary concern of mathematics is that he feels[44]

> ...about number the way Kant felt about space. The positive integers and their arithmetic are presupposed by the very nature of our intelligence, and, we are tempted to believe, by the very nature of intelligence in general.

---
[43][Bishop & Bridges, 1985], p. 4.
[44][Bishop & Bridges, 1985], p. 4.

But, if the positive integers play, for Bishop, the same rôle space plays for Kant, i.e., the positive integers are pure *a priori* conditions of experience, how can Bishop say that mathematics is a creation of the mind? And, secondly, how does Bishop justify the special rôle he attributes to the positive integers?

If we address the second question first, we can argue that, since the exercize of intelligence presupposes the existence of a logic, i.e., of a system of laws of thought, then, if we are prepared to believe that the positive integers have, as Frege and several others think, a logical nature, it follows that the positive integers are, indeed, presupposed by the exercize of human intelligence. As a confirmation of the special rôle of the positive integers, Bishop mentions that[45]

> The development of the theory of the positive integers from the primitive concept of the unit, the concept of adjoining a unit, and the process of mathematical induction carries complete conviction.

Concerning the first question above, Bishop can say that, as in the case of a game of chess there is no conflict between reconciling the pre-existence of the chessmen, the board, and the rules of chess with the creative sequence of actions a game of chess consists of, in the same way there appears to be no *prima facie* conflict between presupposing the existence of a certain intersubjective feature of the human mind and its creative exercize which, in the particular case represented by the positive integers, cristallizes in the development of arithmetic.

In expounding his ideas on the nature and importance of the positive integers within mathematical activity, Bishop distinguishes, very much along verificationist lines, between real and ideal mathematics.

For Bishop:[46]

> ...there are certain mathematical statements that are merely evocative, that make assertions without empirical validity. There are also mathematical statements of immediate empirical validity, which say that certain performable operations will produce certain observable results: for instance, the theorem that every positive integer is the sum of four squares. Mathematics is a mixture of the real and the ideal, sometimes one, sometimes the other, often so presented that it is hard to tell which is which. The realistic component of mathematics — the desire for pragmatic interpretation — supplies the control which determines the course of development and keeps mathematics from lapsing into meaningless formalism. The idealistic component permits simplifications, and opens possibilities which would otherwise be closed. The methods of proof and the objects of investigation have been idealized to form a game, but the actual conduct of the game is ultimately motivated by pragmatic considerations.

---

[45] [Bishop & Bridges, 1985], p. 4.
[46] [Bishop & Bridges, 1985], p. 2.

11. BISHOP'S PHILOSOPHY OF MATHEMATICS 157

The distinction drawn by Bishop between real and ideal mathematics is very similar to the Hilbertian distinction between finitary and ideal mathematics. Indeed, for Hilbert, the finitary fragment of a mathematical theory consists of verifiable results, whereas the ideal fragment allows simplifications and is driven, to say it with Bishop, by 'pragmatic considerations which ultimately refer to the problems of the finitary fragment'.

Furthermore, Bishop's attitude towards idealistic mathematics is in stark contrast with Brouwer's for whom this must be regarded as a collection of pseudo-statements.

Bishop voices his strong disagreement with those who share Brouwer's opinion on idealistic mathematics saying that to believe[47]

> ...that idealistic mathematics is worthless from the constructive point of view ...would be as silly as contending that unrigorous mathematics is worthless from the classical point of view. Every theorem proved with idealistic methods presents a challenge: to find a constructive version, and to give it a constructive proof.

A question we should ask at this point is whether Bishop's constructivism is coherent with his view of idealistic mathematics. In contrast with what some scholars have asserted,[48] there is no incoherence between Bishop's constructivism and his view of idealistic mathematics, because Bishop does not first assert and then deny that non-constructive proofs are compelling. For Bishop, non-constructive proofs are not proofs at all, because they are not procedures of verification. According to Bishop, the value of non-constructive proofs resides exclusively in their heuristic power.

Time has now come to leave behind us the topic of idealist mathematics. In what follows in this section, I shall discuss the aim of Bishop's constructivist programme and compare Bishop's brand of constructivism (BISH) with CLASS, INT and RUSS.

The aim of Bishop's constructivist programme is that of purging mathematical analysis, and the rest of mathematics, of its idealistic content. Bishop intends to do this by (i) developing mathematical theories within a constructivist framework, and (ii) giving 'numerical meaning to as much as possible of' these theories. His motivation for (ii)[49]

> ...is the well-known scandal, expressed by Brouwer (and others) in great detail, that classical mathematics is deficient in numerical meaning.

Now, with regard to the problem of comparing BISH with CLASS, INT, and RUSS, we must notice that Bishop's constructivism[50]

---

[47][Bishop & Bridges, 1985], Prolog, p. 3.
[48][Billinge, 2003].
[49][Bishop & Bridges, 1985], p. 3.
[50][Bridges & Richman, 1988], Ch. 6, p. 120.

> ... is consistent with classical mathematics: every proof in BISH of a statement T is also a classical proof of T. Thus BISH is a generalization of classical mathematics in the same sense that group theory is a generalization of abelian group theory. We pass from BISH to classical mathematics by adopting the law of excluded middle.
>
> [Since] BISH is widely accepted as the common core of INT, RUSS, and any genuinely constructive development of mathematics other than a strictly finitist one ... It follows that, within BISH alone, one cannot disprove anything that is provable in INT or RUSS.

At this point, on the basis of what I have just said about how BISH compares to CLASS, INT, and RUSS, someone might say that the problem concerning the objectivity of mathematics can be successfully solved within a constructive context by BISH, because, if you prove a statement S in BISH then this is a proof also for INT and RUSS.

However, this is not the case for two reasons. First, some proofs in BISH are not accepted by other schools of mathematical constructivism (strict finitism).

Secondly, the objectivity achieved within BISH in relation to RUSS and INT comes at a very high price: BISH is so weak that nothing we can prove within CLASS, RUSS and INT can be disproved within BISH. What this means is that, within BISH, we are unable to distinguish between three mutually incompatible frameworks for mathematics.

This situation is analogous to what we have in geometry with absolute geometry, on one side, and Euclidean and hyperbolic geometries on the other. For, within absolute geometry, it is not possible to choose between the mutually exclusive Euclidean and hyperbolic geometrical systems.

12 Strict finitism

The last variety of constructivist philosophy of mathematics that I am going to discuss in this chapter is the so-called 'strict finitism'. Strict finitism originates from a 'verificationist dissatisfaction' with the traditional constructivist views of mathematics, a dissatisfaction caused by the consideration that these seem to betray verificationist common sense, if there is such a thing.

Indeed, when INT, RUSS and BISH consider mathematical constructions, they always refer to procedures of verification of mathematical statements which can be carried out in principle, i.e., idependently of the resources available to implement such procedures.

'But', the strict finitist asks, 'How can we call "verification" a procedure which in most cases cannot be carried out to completion for lack of resources and which, therefore, in most cases does not produce an answer to the question as to whether a mathematical statement is true or false?'.

12. STRICT FINITISM

As van Dantzig says:[51]

> Unless one is willing to admit fictitious "superior minds" like Laplace's "intelligence", Maxwell's "demon" or Brouwer's "creating subject", it is necessary, in the foundations of mathematics like in other sciences, to take account of the limited possibilities of the human mind [against INT, BISH] and of mechanical devices replacing it [against RUSS].

The strict finitist's criticism of the traditional constructivist approaches to mathematics, besides revealing the presence of an incoherence within the philosophies of mathematics of INT, RUSS and BISH, also shows that the anti-verificationist mythologizing in which INT, RUSS, and BISH engage is not the consequence of their attempt to cope with the difficult task of producing a constructive version of theories such as mathematical analysis, but is something that takes place, as it were, much nearer *home* in elementary number theory.

The classical strict finitist's argument which reveals that INT, RUSS, and BISH 'have a problem', as verificationist frameworks for mathematics, already at the level of elementary number theory is expressed in the following quotation[52]

> Brouwer's "Over de grondslagen der wiskunde" (1907) begins with the words (in translation): "One, two, three ... "; we know this sequence of sounds (spoken ordinal numbers) by heart as a sequence without an end, i.e. continuing itself always according to a known law. If one tries to find out what the dots stand for, one sees that Brouwer's statement can not be maintained. All well-known difficulties of defining the well-ordered transfinite numbers of the second class occur among the spoken ordinal numbers; we do *not* know the "whole" sequence by heart, and it does *not* continue according to a known law. Going on, one arrives at million, ..., billion, ..., trillion, ..., quadrillion, quintillion, sextillion, ... and — knowledge of latin getting scanty — millionnillion, ..., millionnillionnillion, ..., millionilli ...illion (million times repeated), etc., corresponding with
>
> $$\omega, \omega^2, \omega^3, \omega^4, \omega^5, \omega^6, \omega^\omega, \omega^{\omega^\omega}, \omega^{\omega^{\omega^{\cdot^{\cdot^{\cdot}}}}} \; (\omega \text{ times repeated}), \text{ etc.}$$
>
> But: what stands "etc." for?

The strict finitist very clearly shows that once we start rafting on the white waters of verificationism there is no way to stop. The very large finite has, for him, as mythological a nature as the infinite. For, he considers mathematically meaningful only what can be actually verified/constructed, that is, what the resources available to us make possible to verify/construct.

Moreover, van Dantzig argues:[53]

---

[51] [van Dantzig, 1956], p. 273.
[52] [van Dantzig, 1956], p. 276.
[53] [van Dantzig, 1956], p. 273.

Weakening the requirement of actual constructibility by demanding only that one can *imagine* that the construction could actually be performed — or, perhaps one should say rather, that one can *imagine* that one *could* imagine it — means imagining that one would live in a different world, with different physical constants, which might replace the above mentioned upper limit by a higher one, without anyhow solving the fundamental difficulty.

Although strict finitism has to be recognized as the most coherent embodiment of the verificationist standpoint in the philosophy of mathematics, one has to ask whether it can provide a viable philosophy of mathematics.

But how can, under any reasonable interpretation of the term 'viable', a philosophy of mathematics that denies that $10^{10^{10}}$ is a finite number be viable? For, independently of considerations such as 'If n is not a finite number, then $n - 1$ cannot be a finite number either', which paradoxically seem to show that, within a strict finitist framework for number theory, there are no finite numbers, if we were to work simply with very small initial segments of $\mathbb{N}$, we would not be able to study the structure of $\mathbb{N}$ and develop a whole host of results which are then tested (not verified) through their application to other mathematical theories and to the empirical sciences.

From what has been said above, we can safely conclude that, although the strict finitist criticism of INT, RUSS and BISH succeeds in showing that these frameworks for mathematics are incoherent with their commitment to verificationism, and that the problem of what we must understand by 'constructive proof in mathematics' is as open as ever, the strict finitist's obsession with actual procedures of verification, rather than bringing about a much desired reconciliation between mathematical ontology and mathematical epistemology, leads to paradox and, therefore, to mathematical nihilism.

However, independently of whether or not strict finitism provides a tenable philosophy of mathematics, it is undeniable, from what we have seen so far of the various constructivist schools of thought, that there exists a profound and irreconcilable disagreement among constructivists about what a constructive proof in mathematics is.

The disagreement existing among constructivists concerning the notion of constructive proof is profound, because it involves 'the' fundamental notion of mathematical constructivism; and is irreconcilable, because, in contrast with what would happen in a debate among, say, biologists on the concept of species, no matter of fact about constructive proofs — which determines the correctness/incorrectness of the various proposals put forward — can be presupposed by the constructivists, lest they turn out to be realists about constructive proofs.

One of the most serious consequences of this state of affair is that mathematical constructivism cannot save mathematical objectivity and, therefore, cannot provide a tenable philosophy of mathematics either.

CHAPTER 5

THE THIRD WAY: A REALISM WITH THE HUMAN FACE

1 Introduction

In what follows questions (a) 'What is the nature of mathematical reality?' and (b) 'How do we acquire mathematical knowledge?' will be addressed while outlining a realist philosophy of mathematics which commits one neither to dream the dreams of Platonism nor to reduce the word 'realism' to mere noise. It will be argued that mathematics is a science of patterns, where patterns are neither objects nor properties of objects, but aspects, or aspects of aspects, etc. of concrete objects, which become perspicuous to the observer when a system of representation is in place (answer to question (a)).

This view will strike a middle path between Platonism and anti-realism. In fact, according to it patterns are real, because they ultimately depend on concrete objects, but are neither conceivable as objects nor as properties of objects, because they also depend on a system of representation.

It is the characteristic of mathematical patterns of being dependent on a system of representation what makes it possible to avoid Platonism, and gives origin to a 'realism with the human face'. It makes it possible to avoid Platonism, because since patterns are neither objects nor properties of objects, they cannot *a fortiori* be abstract objects. And it gives origin to a realism with the human face, because a pattern, very much like a Kantian phenomenon, is the outcome, among other things, of the ordering function performed on perception (and other types of representation) by the man-made systems of representation.

It will also be argued that the ontology of mathematical patterns admits of a very natural explanation of how we acquire knowledge of them (answer to question (b)).

2 Patterns

In the philosophical literature the word 'pattern' has been used in a variety of ways. It is, therefore, extremely important, at the very beginning of this discussion, to clarify what I mean by it.

What I mean by 'pattern' is an *aspect* of a concrete object (or an aspect of an aspect of a concrete object, etc.), which becomes perspicuous to us when we consider the object (or the aspect of ...) in relation to a given mathematical theory.

For example, if writing $X = \{a, b, c\}$ is a way of seeing the chairs in a room as elements of a set, we could also see these very chairs as individual solids and proceed to comparing them with one another in relation to properties of their surface.

Here the notion of representation plays an extremely important rôle. Seeing the chairs in a room as the elements of a set $X = \{a, b, c\}$ is giving a representation, within the theory of sets, of the objects observed. But since we could not see a group of chairs as a set, if we did not have the concept of set, seeing something as, and, in particular, seeing something as mathematically meaningful, presupposes the existence of a system of representation. As Wittgenstein has remarked in the *Philosophical Investigations*:[1]

> It is only if someone *can do*, has learnt, is master of, such-and-such, that it makes sense to say he has had *this* experience. And if this sounds crazy, you need to reflect that the *concept* of seeing is modified here. (A similar consideration is often necessary to get rid of a feeling of dizziness in mathematics.)

This, of course, does not mean that objects are *naked* in the sense that their aspects are simply a function of the theory, because only one of the possible representations that we can give within the chosen theory is faithful.[2] To stick to the example I have given above, we could not be right in saying that the set of chairs in the room contains 4 elements.

The fact that there is only one faithful representation of those objects within a given theory T shows that aspects are *real*, but the fact that an aspect is theory-dependent, in the sense that it depends on the particular theory T chosen, whereas the object does not, shows that aspects are not objects.

Within this view of patterns, we can have patterns of patterns. In fact, if our set $X = \{a, b, c\}$ expresses the set-theoretical pattern of the chairs in a room, we can generate from X the set $\mathcal{P}(X)$, the power set of X. Now $\mathcal{P}(X)$ can be seen as a lattice, when partially ordered by inclusion, or as the discrete topology τ on X, when we consider the elements of $\mathcal{P}(X)$ as open sets. What is interesting in this example is that, given $\mathcal{P}(X)$, if we study it through the theory of partial orders, then a particular aspect, or

---

[1] See [Wittgenstein, 1983], Part II, §xi, p. 209^e.

[2] A *representation* of an object is a description of the object expressed by propositions. A given representation of an object is *faithful* just in case the proposition (or propositions) which express it is (are) true.

3. OBJECTS AND PROPERTIES OF OBJECTS

pattern, of $\mathcal{P}(X)$ is revealed, i.e., $\mathcal{P}(X)$ is a lattice. If, instead, we consider $\mathcal{P}(X)$ through topologically tinted glasses, we will discover another aspect of $\mathcal{P}(X)$, i.e., the fact that it constitutes the discrete topology on X.

But if patterns are aspects of concrete objects (or aspects of aspects of concrete objects, etc.) which become perspicuous to us when we consider the objects (or the aspects of ...) in relation to a given mathematical theory, how do we distinguish between mathematical and non-mathematical aspects? The distinction between mathematical and non-mathematical aspects is determined by the difference existing between mathematical and non-mathematical systems of representation. We could see the same object in one case as a sphere and in the other as a snooker ball according to which system (mathematical/non-mathematical) of representation we adopt.

However, if patterns are real but are not objects, what are they? Are they properties of objects?

3 Objects and properties of objects

If by 'object' we mean what exists independently of whether we are thinking about it or not, it follows, as we have seen in the previous section, that patterns are not objects. But could patterns be properties of objects?

To clarify this question, let us imagine that in a given room $\mathcal{R}$ there are only red chairs, and that it is true to say that 'There are three chairs in $\mathcal{R}$'. We could then correctly assert that 'There are so-many undivided chair-parts in $\mathcal{R}$', for 'so-many' different from three, etc.

Hence, what we numerically see the content of the room as varies according to which system of representation we adopt: in one system, the unit of representation being *chair*, we would describe the content of $\mathcal{R}$ as 'three chairs'; in the other, the unit of representation being *undivided chair-part*, we would describe the content of the room as 'so-many undivided chair-parts'. Therefore, we would here have, as Frege remarked in one of his objections against Mill,[3]

> ... a difference in number to which no physical difference corresponds ...

This phenomenon, according to which to the dawning of a mathematical aspect there corresponds no physical property in the object(s) perceived, is not peculiar to number-theoretical aspects, but is common to all mathematical aspects that dawn on us. For instance, when we see something as a triangle, or as a circle, or as a continuous function, the obtaining of the phenomenon mentioned above is particularly clear, because none of the objects perceived is, as a matter of fact, a perfect triangle or a perfect circle, or a continuous entity.

---

[3][Frege, 1884], §25, p. 33.

On the other hand, in our attempt to describe the content of $\mathcal{R}$ we make also use of concepts which individuate properties such as 'the total portion of space-time occupied by the objects in $\mathcal{R}$', 'the colour and shape of the objects in $\mathcal{R}$', etc. which appear to remain invariant under change of systems of representation and to which there corresponds a physical difference, i.e., if they were different then also the objects perceived would be physically different. The invariance of these latter properties under change of systems of representation and their *physical* relevance justify us in considering them as properties of the content of $\mathcal{R}$, and in distinguishing them from what we have called 'mathematical patterns'.[4]

4 External and internal relations

If patterns are neither objects nor properties of objects, what are they? Perhaps the following quotation from Wittgenstein's *Philosophical Investigations* will shed some light on this question:[5]

> The colour of the visual impression corresponds to the colour of the object (this blotting paper looks pink to me, and is pink) — the shape of the visual impression to the shape of the object (it looks rectangular to me, and is rectangular) — but what I perceive in the dawning of an aspect is not a property of the object, but an internal relation between it and other objects.

Given the quotation above, the questions that we need to ask are 'What is, for Wittgenstein, an internal relation?' and 'What is the relevance of such a notion to our concept of pattern?'

In the *Philosophical Investigations* Wittgenstein does not say anything about the nature of internal relations and his silence might be evidence in favour of the opinion that his views on the matter held in the *Tractatus*,[6] the *Notebooks*[7] and the *Notes on Logic*[8] remained, to a large extent, unchanged in his later thought.

What I take these views to be can be summed up in the two following theses: i) a 2-place relation R is *internal* to two objects a and b just in case it is impossible (Wittgenstein says 'unthinkable') that a and b do not stand in relation R to one another, ii) internal relations (and properties) are what determines the features, structure of a fact.[9]

---

[4]The considerations in the main text should put to rest the doubts expressed in [Ernest, 1999] concerning

> ...the assumption that we can know unproblematically whether a pattern is an aspect of an object, relative to a given theory.

[5][Wittgenstein, 1983], Part II, p. 212^e.
[6]See [Wittgenstein, 1981].
[7]See [Wittgenstein, 1979a].
[8]See [Wittgenstein, 1979b].
[9][Wittgenstein, 1981], §4.1221, p. 27:

4. EXTERNAL AND INTERNAL RELATIONS

One of the most important consequences of characteristics i) and ii), within the system of the *Tractatus*, was that internal relations performed a rôle typical of elements of what Wittgenstein called 'logical form', that is, what he thought propositions and reality must have in common for representations of reality to be possible.[10]

The changes occurred in the later period were mainly related to the metaphysical account of the *formality* of internal relations. What I mean by this is that, although internal relations were still seen by Wittgenstein in the later period not as properties of objects, but as what was part of our system of representation of objects, he had abandoned his early period account of how such representations are possible.

However, this is not the place to carry out Wittgensteinian exegesis, but to produce clarifications concerning the concept of internal relation.

As in other occasions, Wittgenstein's ideas, despite their suggestiveness, do not produce sharp characterisations of the notions analysed. I find, on the contrary, that G. E. Moore's way of tackling the problem of internal relations, besides capturing characteristics i) and ii) of Wittgenstein's remarks on internal relations, provides us with a clear and informative definition.

G. E. Moore, in his seminal paper 'External and Internal Relations',[11] produces an application of the modal context to give a characterization of internal relations able to distinguish them from external relations.[12]

An internal property of a fact can also be called a feature of that fact (in the sense in which we speak of facial features, for example).

and ibid., §4.123:

A property is internal if it is unthinkable that its object should not possess it. (This shade of blue and that one stand, eo ipso, in the internal relation of lighter to darker. It is unthinkable that *these* two objects should not stand in this relation.) (Here the shifting use of the word 'object' corresponds to the shifting use of the words 'property' and 'relation'.)

[10][Wittgenstein, 1981], §4.122, pp. 26-27:

In a certain sense we can talk about formal properties of objects and states of affairs, or, in the case of facts, about structural properties: and in the same sense about formal relations and structural relations. (Instead of 'structural property' I also say 'internal property'; instead of 'structural relation', 'internal relation'. I introduce these expressions in order to indicate the source of the confusion between internal relations and relations proper (external relations), which is very widespread among philosophers.) It is impossible, however, to assert by means of propositions that such internal properties and relations obtain: rather, this makes itself manifest in the propositions that represent the relevant states of affairs and are concerned with the relevant objects.

[11]See [Baldwin, 1993], pp. 79-105.
[12]Moore's methodology in tackling this problem anticipates of many years that adopted by Kripke in [Kripke, 1980].

Before I begin a discussion of Moore's definition, I must say that Moore does not deal with relations in general, but with what he calls 'relational properties'.[13] Having made these provisos, which do not alter the relevance or the generality[14] of the discussion, I can give Moore's definition of internal relational property:[15]

> Let Φ be a relational property, and A a term to which it does in fact belong. I propose to define what is meant by saying that Φ is internal to A ... as meaning that from the proposition that a thing has not got Φ, it 'follows' that it is *other* than A.

To express this in modern terminology we can say that a relational property Φ is internal to a term A just in case A has the property Φ and it is *necessary* that for any x, if x does not have the property Φ then x is different from A.[16]

An example of internal relational property is the following. Let A and B be triangles in a Euclidean plane α and $\Phi(x) :=$ 'x has the same number of angles as B' a relational property. Then clearly Φ is internal to A.

An example of external relational property may be the following. Let $(\mathbb{Z}, +)$ be the algebraic structure obtained when we define $+$ on $\mathbb{Z}$ (a group); $A := 2$; and $\Phi'(x) :=$ 'x is the inverse of -2'. In this case we have that $\Phi'(A)$ is true, because 2 is the additive inverse of -2, but, if we consider the structure $(\mathbb{Q} \setminus \{0\}, \times)$ then 2, in this case, is *not* the inverse of -2.[17]

---

[13] If $x < y$ is a 2-place relation, which has as extension the set $B = \{(a,b) : a, b \in \mathbb{R} \text{ and } a < b\}$ ($B \subseteq \mathbb{R}^2$), we call $x < 5$ 'relational property of real numbers', because the extension of $x < 5$ is the set $A = \{x : x \in \mathbb{R} \text{ and } x < 5\}$, (in this case $A \subseteq \mathbb{R}$).

[14] If you can obtain a relational property from a 2-place relation by substituting an individual constant for a free variable, you can obtain a 2-place relation from a relational property by substituting for an individual constant a variable which differs from the variables (free or bound) occurring in the expression.

[15] See [Moore, 1922], p. 90.

[16] Formally speaking we could put it in this way:

$$(\Phi(A) \land \Box \forall x(\neg \Phi(x) \to \neg(x = A))).$$

[17] This means that

$$\Diamond \exists x(\neg \Phi'(x) \land x = A)$$

is true and, therefore,

$$\Box \forall x(\neg \Phi'(x) \to \neg(x = A))$$

is false, therefore, Φ' is an external relational property. Moore, in the same paper, draws also the conclusion that the propositions formalised by

(*) $(\Phi(A) \land \Box \forall x(\neg \Phi(x) \to \neg(x = A))$

and

(**) $(\Phi(A) \land \forall x \Box(\neg \Phi(x) \to \neg(x = A))$

are not equivalent. Indeed, when you consider external relational properties, (**) is true and (*) is false.

4. EXTERNAL AND INTERNAL RELATIONS

From the Moorian definition of the notion of internal relational property (and, therefore, also from the corresponding definition of internal relation), we can show how this satisfies condition i) of Wittgenstein's requirements by saying that if Φ is an internal relational property of an object A then it is *necessary* for A to have such a property or, equivalently, it is *impossible* to have A without $\Phi(A)$ being true (Wittgensteinian condition).

The second Wittgensteinian condition (formality) that must be satisfied by an internal relation can also be seen to obtain for internal relational properties in Moore's sense. In fact, since an internal relational property of an object A is a property without which A could not exist, and since A has also many external relational properties which are important to characterise it as an individual, the set of internal relational properties has as elements the necessary conditions for A to be individuated and these conditions, if they are not also sufficient, can only individuate the general form that something must have if it has to be an A.

Having so clarified the notion of internal relational property (and, therefore, that of internal relation), we must now turn to the problem of checking the claim made that seeing a mathematical pattern can be explained as the dawning on us of an internal relation holding between objects.

It seems to me that Wittgenstein's account of the dawning of an aspect contains two extremely important positive features and two negative ones. The positive features are that in the dawning of an aspect (1) we do not perceive a property of an object; (2) what we perceive is a relation between objects. The negative features are that (3) the relation we perceive in the dawning of an aspect is an internal relation; and that (4) such an internal relation is between the object in question and other objects. Since I have already argued in favour of points (1) and (2), what is left for me to do is providing evidence against points (3) and (4).

The example of external relation I gave above will do very well as a counterexample for (3). In fact, although given a particular structure $(\mathbb{Z}, +)$, the dawning of an aspect, which consists in 'seeing 2 as the inverse of -2' in $(\mathbb{Z}, +)$, is not a property of 2, etc. but is a relation between 2 and -2, such a relation is *not* internal, because there are structures, e.g., $(\mathbb{Q}\setminus\{0\}, \times)$ in which it is false to assert that '2 is the inverse of -2'.

Secondly, the idea that when 'we see an object a as ...' what we perceive is a relation between a and other objects which constitute some kind of context, or background, against which a is perceived as ... seems to exclude the case in which the aspect of a which dawns on us is not the consequence of a relation that we perceive between a and other objects, but rather the consequence of seeing a as composed of parts which are related to each other in such-and-such a way.

An example of this phenomenon is given by seeing an object a as a circle. Such a perception is made possible by our system of representation containing concepts such as 'point', 'distance', 'centre', etc. and by our seeing the component parts of a being related to each other in such a way that the set of border points of a are perceived to have the same distance from the centre point of a.

Therefore, although Wittgenstein's account of the dawning of an aspect contributes to revealing the structural nature of what is investigated by mathematics and opens, as we shall see in the final sections of this chapter, the door for the construction of a plausible epistemology of mathematics, it turns out to be too restrictive. But perhaps, before bringing this section to a close, I ought to clarify in what sense Wittgenstein's account of the dawning of an aspect contributes to revealing the structural nature of what is investigated by mathematics.

If mathematics is a science of patterns and patterns are, as Wittgenstein argues, neither objects nor properties of objects, but internal (or external) relations defined on a certain domain, that is, on the set of parts of an object a, or of a pattern Π, or of ... it follows that the pair $(\mathcal{O}, \Pi)$, formed by a pattern Π and by the set of entities on which Π is defined, is a structure according to the definition I gave in §3.10; and that, consequently, the view according to which mathematics is a science of patterns, where patterns are defined as above, is a version of structural realism.

It is important to consider that this is a form of structural realism which, like any type of structural realism, has the advantage of being in agreement with the present day development of mathematics, but which, unlike any other type of structural realism (Resnik's, Shapiro's, etc.), is *not* a form of Platonism about structure.

In fact, on the one hand, contemporary mathematics is more and more a science of abstract structures — what else are groups, rings, fields, topological spaces, etc.? — even though the term 'structure' in mathematics refers to different types of entities. And, on the other hand, being neither objects nor properties of objects, patterns are not conceivable as abstract mathematical entities which exist independently of thinking about them.

However, at this point someone might say 'If by "particular" we mean what occupies a unique portion of space-time', and is such that 'All identifying description of [it] may include, ultimately, a demonstrative element',[18] we have to conclude that patterns are not particulars. That this is, indeed, the case is shown by the fact that we can, for example, see various different plates laid on the same table as circles.

---

[18][Strawson, 1979], Part I, Ch. 1, [2], p. 22.

5. NEITHER IN RE NOR ANTE REM STRUCTURES

On the other hand, if patterns are not particulars, they must be universals, and these are universals which, as we know from the previous discussion, do not exist independently of thinking about them. But if this is the case, how do we characterize their reality?'

5 Neither *in re* nor *ante rem* structures

Having established that patterns are real, but they are universals too, we face the classical choice: are they *in re* or *ante rem*?

Unfortunately, this classical dichotomy is not able to sort out the question concerning the reality of mathematical patterns. For a mathematical pattern, in contrast with other types of universals, is neither *in re* nor *ante rem*. It is not *in re*, because, apart from the consideration that if you change system of representation then also what you see something as changes, we have that concrete objects are *not* (perfect) circles, triangles, etc. And it is not *ante rem* either, because a mathematical pattern does not exist independently of the object(s) (or aspects, etc.) we see as ...

Therefore, taking into account the peculiarities of patterns, we can say that a pattern is the structure (form) of a representation of an object (or of a pattern, etc.).

When we represent an object (or a pattern), and this representation takes place in perceptual space, we must distinguish, as Kant showed, between two important elements of such representations (1) the sensory input, that is, colours, sounds, smells, etc.; and (2) the form (or structure) of the representation itself, that is, what turns sensory input into a pre-reflectively ordered manifold. The pattern that dawns on us — what we see something as — is the structure (or form) of our representation.

It is very important to remark on the fact that the distinction between the form and the content of a representation applies both to perceptual and non-perceptual representations alike. But, if it is clear what a perceptual representation is, it is perhaps not so clear what is meant by 'non-perceptual representation'.

A non-perceptual representation is a representation which takes place in linguistic rather than in perceptual space. An example of non-perceptual representation is given by the algebraic description of a curve: a circle in $\mathbb{R}^2$ with centre (a, b) and radius r can be described (represented) as the set of solutions of the equation

$$(x - a)^2 + (y - b)^2 = r^2.$$

Although we perceive the symbols which are the component parts of the equation which represents an aspect of a certain object, the symbols are not perceptions, but elements of the alphabet of a language in relation to which the equation above is a well-formed formula.

In the particular example of non-perceptual representations mentioned above, the content of our representation is given by the points over which x and y range and to which a and b refer, whereas the pattern (the way these points are related to each other) is what is expressed by the equation.

The idea of a pattern as the structure (form) of a representation of an object (or of a pattern) has a distinctive Kantian flavour to it in that it speaks of the reality and structure of phenomena as entities which are dependent both on objects and on systems of representation.

However, there also are a number of important differences between the views on perception defended in this book and Kant's. Such differences are, in particular, related to the rôle played by concepts belonging to systems of representation in structuring representations in a pre-reflective way (see §§5.8-5.9).

But now, since patterns in general, and mathematical patterns in particular, have been here defined as forms of representation of objects (or of patterns, etc.), someone might wonder whether a pattern is some kind of Husserlian essence.

If we consider that, for Husserl, the essence of an individuum is the component of the being of that individuum which determines the characteristics that enable us to identify it and distinguish it from other individua (see §3.8), we must then conclude that patterns are not essences in Husserl's sense.

In fact, seeing the Jastrow duck-rabbit figure as a duck is so far from unveiling to us 'the component of the being of that individuum etc.' that we can see the same individuum as a rabbit. The same considerations, of course, apply to the mathematical case, in which, for example, we can see the same object in one case as a cube and in the other as a distributive lattice switching at will between the two ways of representing it.

Secondly, if it is correct what I say in §3.8 about the strong similarity existing between Husserl's realm of essences and Frege's Third Realm, it follows that patterns are not Husserlian essences. For, although universals, they are not conceivable as elements of a reality which exists independently of the existence of concrete objects, and independently of that part of human activity which results in the construction of systems of representation.

There is, indeed, a whole cumulative hierarchy of mathematical patterns which has at its bottom level patterns which are aspects of concrete objects, at the second level the elements of the bottom level plus the patterns of patterns belonging to the bottom level, etc. Of course, the elements of a level of the hierarchy indexed by a limit ordinal α are the elements of the union of the levels of the hierarchy which are indexed by an ordinal β, for $\beta < \alpha$.

6 Which realism?

My statement that patterns are real, but are neither objects nor properties of objects, might lead some readers into thinking that the type of realism I am advocating is some kind of *internal realism* in Putnam's sense. It is to try to clarify the difference between Putnam's views and mine that I have decided to write the present section.

Internal realism, as I understand it, is the view according to which i) the answers to questions such as 'How many objects are there in this room?' are relative to the conceptual schemata adopted,[19] ii) there is nothing more to truth than warranted assertability.[20]

Independently of the question whether what Putnam calls 'internal realism' is realism at all, it is extremely important to realise that in my discussion of the reality of patterns I neither question the existence of objects nor give any assent to the thesis postulating the relativity of these to a given conceptual schema. What I do is: assuming that there are objects, that is, entities which exist independently of anybody's thinking about them, attempt to show that what I call 'patterns' are neither objects nor properties of objects.

However, turning for a moment the attention to the question of *how* we should talk about objects, it seems to me that there is little doubt that *what* I draw below (Fig. 5.1) exists independently of whether or not I am thinking about it.

---

[19]See [Putnam, 1992], Ch. 7, p. 113–114:

...there are many ways of *using* the notion of an object — even the so-called "logical notion" of an object — or the existential quantifier. And, depending on how we use the notion, the answer to the question "How many objects are there in the room?" can be "Five", "Seven", "2^n" — and there are many more possibilities. A metaphor which is often employed to explain this is the metaphor of the cookie cutter. The things independent of all conceptual choices are the dough; our conceptual contribution is the shape of the cookie cutter. Unfortunately, this metaphor is of no real assistance in understanding the phenomenon of conceptual relativity. Take it seriously, and you are at once forced to answer the question 'What are the various parts of the dough?' If you answer that (in the present case) the 'atoms' of the dough are the n elementary particles and the other parts are the mereological sums containing more than one 'atom', then you have simply adopted one particular transcendental metaphysical picture: the picture according to which mereological sums 'really exist'.

[20][Putnam, 1992], p. 115:

the suggestion which constitutes the essence of "internal realism" is that truth does not transcend use.

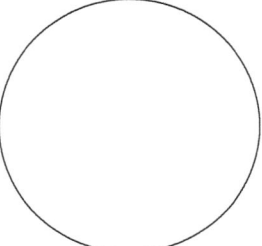

Figure 5.1.

This in the sense that, given the present state of our knowledge, it is unreasonable to suppose that my thinking, or the lack of it, about what I have drawn has any effect whatsoever on the mark left by the ink on the paper.

On the other hand, it seems to me that there is also little doubt that the aspect — numerical, geometrical, etc. — that dawns on me when *I see it as* ... is, as I argued in this book, doubly dependent on the object and on the system of representation. To this Putnam might reply 'But you have already decided in your system of representation what the objects are, i.e., points, lines, etc., and how you are going to manipulate the symbols which refer to these objects'. The answer to this is in the purest Kantian style, namely, in my system of representation I have simply defined the features in terms of which I am going to describe what I have drawn.

My system of representation provides only the *negative* condition for knowledge, the positive condition being how things are, which is something that can be determined only by experience.[21]

---

[21] See [Kant, 1787], I Transcendental Doctrine of Elements, Second Part. Transcendental Logic, §III, pp. 97–98:

> The question, famed of old, by which logicians were supposed to be driven into a corner, obliged either to have recourse to a pitiful sophism, or to confess their ignorance and consequently the emptiness of their whole art, is the question: What is truth? The nominal definition of truth, that it is the agreement of knowledge with its object, is assumed as granted; ... as regards knowledge in respect of its mere form (leaving aside all content), it is evident that logic, in so far as it expounds the universal and necessary rules of the understanding, must in these rules furnish criteria of truth. Whatever contradicts these rules is false. For the understanding would thereby be made to contradict its own general rules of thought, and so to contradict itself. These criteria, however, concern only the form of truth, that is, of thought in general; and in so far they are quite correct, but are not by themselves sufficient. For although our knowledge may be in complete accordance with logical demands, that is, may not contradict itself, it is still possible that it may be in contradiction with its object. The purely logical criterion of truth, namely, the

6. WHICH REALISM?

Therefore, if this is correct, not only do objects exist independently of conceptual schemata, but the truth of statements such as those expressing descriptions of what I have drawn cannot be characterised simply as warranted assertability.

In talking about objects the way I do I am not introducing a mythology of a thing in itself, a noumenal entity — what Putnam caricatures with the metaphor of the *dough* (see footnote 19) — lying *behind* the veil of aspects that we perceive. An object is what generates the *a posteriori* consistency and mutual informativity of the correct descriptions of the aspects that we perceive of it.

If it is clear how the existence of an object might determine the mutual informativity of the correct descriptions of the aspects that we perceive of it, perhaps it is necessary to spend a few words to explain why an object determines the *a posteriori* consistency of correct descriptions of its aspects. I can give an example illustrating this point in the following way. Assume that there is a substance S which can be used as ink to write on several materials and that there is a material M which can be used to be written upon with several types of ink. Assume also that there are some books written with ink S and others using material M.

However, when we attempt to write using ink S on material M a chemical reaction takes place that makes hopelessly illegible whatever we write. If this is the case, we will never have a book written with ink S on material M and this is not something that we can determine *a priori*, this is something that we discover empirically through chemical investigations of how S and M react when combined together.

Therefore, if we correctly describe something as being a book then this cannot be what has been obtained using ink S on material M and the aspects 'Book written with ink S' and 'Book written on material M' are not compatible as a consequence of the properties of the object.

Returning now to the main problem of this section — whether or not objects are relative to conceptual schemata — when I see what occupies a certain portion of space-time 'as a table', 'as an undivided collection of table-parts', 'as a collection of k molecules', etc. and my descriptions are correct I am seeing aspects of what occupies that portion of space-time. The reason for saying this lies in that i) the different ways in which I perceive the content of the given portion of space-time are not inconsistent with one another, on the contrary, ii) they increase the amount of information

---

agreement of knowledge with the general and formal laws of the understanding and reason, is a *conditio sine qua non*, and is therefore the negative condition of all truth. But further than this logic cannot go. It has no touchstone for the discovery of such error as concerns not the form but the content.

I possess about it, but not in a simply additive way, in that iii) this is information I can pass on to and agree with other people about and in that iv) by simply varying the conceptual schema adopted I can switch from *seeing the content of the room as...* to *seeing the content of the room as ...something else* in conditions where there is no reason to suppose that the perceptual input has changed.

Putnam is impressed by the fact that when we ask 'How many objects are there in this room?', we may be able to give and justify several different replies which are all true, but which seem to be inconsistent with one another, e.g., '5', '7', etc. To explain this strange phenomenon he flies into the arms of the relativist saying that the reply we give to the question 'How many objects are there in this room?' essentially depends, as those who mock philosophers usually say, on 'What you mean by "object" '.

But, *pace* Putnam, Quine and all believers in inscrutability of reference, there is no need to do this, if we interpret the question above as asking 'How do you numerically describe the content of this room?'. What you refer to when you use the definite description 'The content of this room' exists independently of any conceptual schema. It is the particular way adopted of describing it numerically that changes.

7 Kant on concepts and perception

So far in this chapter I have explored a few questions concerning the ontology of mathematical patterns. The answers given to these questions ought to have produced an explanation of the reason why the position defended in this book can be correctly described as structural realism, and an elucidation of the profound difference existing between this type of realism and what is known as Platonism about structures (see §3.10).

However, there is, of course, another cluster of problems revolving around the attempt to give an account of the phenomenon known as 'seeing an aspect'. Such an account is, of course, very important to produce a satisfactory epistemology of mathematics.

In what I have called 'seeing something as', it seems that an interesting phenomenon takes place, this is the interaction between concepts and perceptions. Before attempting to clarify what sort of relation between concepts and perceptions is revealed by aspect seeing, it is a good idea to expound briefly the traditional Kantian doctrine about these matters.

For Kant, the rôle of concepts is that of producing understanding by means of the activity of *combination* that they perform over the manifolds independently given (independently of concepts) through intuitions.[22]

---

[22]See [Kant, 1787], Transcendental Analytic, Section 2 Transcendental Deduction of the Pure Concepts of Understanding, §15, p. 151:

7. KANT ON CONCEPTS AND PERCEPTION   177

(What Kant means by intuition is 'That representation which can be given prior to all thought ...'.)[23]

This feature of Kant's ideas about the sharp separation existing in the way that human reason operates between perception and concepts emerges very strongly from the following passage in which Kant compares human understanding with Divine understanding:[24]

> ... were I to think an understanding which is itself intuitive (as, for example, a divine understanding which should not represent to itself given objects, but through whose representation the objects should themselves be given or produced), the categories would have no meaning whatsoever in respect of such a mode of knowledge. They are merely rules for an understanding whose whole power consists in thought, consists, that is, in the act whereby it brings the synthesis of a manifold, given to it from elsewhere in intuition, to the unity of apperception — a faculty, therefore, which by itself knows nothing whatsoever, but merely combines and arranges the material of knowledge, that is, the intuition, which must be given to it by the object.

If my interpretation is correct, for Kant, concepts do not play any rôle in perception, whose sole factors are the objects belonging to the external world, the senses and the *a priori* pure intuitions of space and time. Once perception takes place, concepts represent the conditions according to which human understanding can be generated.

However, it is extremely important to point out that even though concepts, for Kant, have no part in perception they are by him considered as *a priori* conditions of experience:[25]

> Now there are two conditions under which alone the knowledge of an object is possible, first, *intuition*, through which it is given, though only as appearance; secondly, *concept*, through which an object is thought corresponding to this intuition. It is evident from the above that the first condition, namely, that under which alone objects can be intuited, does actually lie *a priori* in the mind as the formal ground of the objects. All appearances necessarily agree with this formal condition of sensibility, since only through it can they appear, that is, be empirically intuited and given. The question now arises whether *a priori* concepts do not also serve as antecedent conditions under which alone anything can be, if not intuited, yet thought as object

---

The manifold of representations can be given in an intuition which is purely sensible, that is, nothing but receptivity; and the form of this intuition can lie *a priori* in our faculty of representation, without being anything more than the mode in which the subject is affected. But the combination (*conjunctio*) of a manifold in general can never come to us through the senses, and cannot, therefore, be already contained in the pure form of sensible intuition. For it is an act of spontaneity of the faculty of representation; and since this faculty, to distinguish it from sensibility, must be entitled understanding, all combination — be we conscious of it or not, be it a combination of the manifold of intuition, empirical or non-empirical, or of various concepts — is an act of the understanding.

[23] See [Kant, 1787], ibid., §16, p. 153.
[24] See [Kant, 1787], ibid., §21, p. 161.
[25] See [Kant, 1787], ibid., Section I, §14, p. 126.

in general. In that case all empirical knowledge of objects would necessarily conform to such concepts, because only as thus presupposing them is anything possible as *object of experience*. Now all experience does indeed contain, in addition to the intuition of the senses through which something is given, a *concept* of an object as being thereby given, that is to say, as appearing. Concepts of objects in general thus underlie all empirical knowledge as its *a priori* conditions. The objective validity of the categories as *a priori* concepts rests, therefore, on the fact that, so far as the form of thought is concerned, through them alone does experience become possible. They relate of necessity and *a priori* to objects of experience, for the reason that only by means of them can any object whatsoever of experience be thought.

There is no contradiction here generated by the opinions expressed above concerning the function of concepts, because, for Kant, there is much more to experience than mere perception:[26]

Experience is an empirical knowledge, that is, a knowledge which determines an object through perceptions. It is a synthesis of perceptions, not contained in perception but itself containing in one consciousness the synthetic unity of the manifold of perceptions. This synthetic unity constitutes the essential in any knowledge of *objects* of the senses, that is, in experience as distinguished from mere intuition or sensation of the senses. In experience, however, perceptions come together only in accidental order, so that no necessity determining their connection is or can be revealed in the perceptions themselves. For apprehension is only a placing together of the manifold of empirical intuition; and we can find in it no representation of any necessity which determines the appearances thus combined to have connected existence in space and time. But since experience is a knowledge of objects through perceptions, the relation [involved] in the existence of the manifold has to be represented in experience, not as it comes to be constructed in time but as it exists objectively in time. Since time, however, cannot itself be perceived, the determination of the existence of objects in time can take place only through their relation in time in general, and therefore only through concepts that connect them *a priori*. Since these always carry necessity with them, it follows that experience is only possible through a representation of necessary connection of perceptions.

These ideas of Kant remained very influential for quite some time. Many analytical philosophers took on board the Kantian tenet that concepts have nothing to do with perception reformulating it in a more modern terminology. An example of this is given by the traditional distinction operated within analytical philosophy between sentences that are *observational* and sentences that are not.

The factor that is common to the several definitions provided by different philosophers of what an observation sentence is is that 'A sentence S is *observational* just in case S is not theory-laden'. The sentence 'X is a microscope' is theory-laden, because to understand what a microscope is we need to know optics, etc., therefore, observation sentences ought to be sentences that we can understand independently of any theory. In a sense observation sentences ought to be reports of our observation rather than of

---

[26] See [Kant, 1787], ibid., Book II, Ch. II, 3 Analogies of Experience, pp. 208–209.

our interpretation, reports about what we see and not reports about how we interpret what we see.

Such a Kantian distinction between observation and non-observation sentences has been attacked from many quarters. It has been, in particular, attacked by philosophers of science such as Kuhn who believe that there are no basic facts which are observed/perceived by everybody in the same way, but that the adoption of a particular scientific theory affects also what we perceive:[27]

> Since remote antiquity most people have seen one or another heavy body swinging back and forth on a string or chain until it finally comes to rest. To the Aristotelians, who believed that a heavy body is moved by its own nature from a higher position to a state of natural rest at a lower one, the swinging body was simply falling with difficulty. Constrained by the chain, it could achieve rest at its low point only after a tortuous motion and a considerable time. Galileo, on the other hand, looking at the swinging body, saw a pendulum, a body that almost succeeded in repeating the same motion over and over again ad infinitum. And having seen that much, Galileo observed other properties of the pendulum as well and constructed many of the most significant and original parts of his new dynamics around them. From the properties of the pendulum, for example, Galileo derived his only full and sound arguments for the independence of weight and rate of fall, as well as for the relationship between vertical height and terminal velocity of motions down inclined planes. All these natural phenomena he saw differently from the way they had been seen before.

However, if Kuhn, and others, succeeded in showing that there are no such things as observation sentences they did not provide us with a new theory of experience which might recognise and give an account of the rôle played by concepts in perception. This latter problem is, in my view, at the heart of the second part of Wittgenstein's *Philosophical Investigations* and will be the object of analysis of the following section.

8 Seeing or interpreting?

Budd, in his interesting analysis of Wittgenstein's position on *aspect seeing* reaches the conclusion that, for Wittgenstein:[28]

> ...the concept of seeing an aspect *lies between* the concept of seeing colour or shape and the concept of interpreting: it resembles both of these concepts, but in different respects.

The concept of seeing an aspect, says Budd, is similar to that of seeing a colour, because when we say that we see something as a cube or as a wire-box it does not make any sense to ask whether what we are seeing is true or false. We are not making a conjecture, we are having a perception and reporting it. In other words, there is no being right or wrong involved here.

---
[27]See [Kuhn, 1970], Ch. X, pp. 118–119.
[28]See [Budd, 1987], p. 16.

Moreover, 'seeing an aspect' and 'seeing a colour' are terms that refer to *states*. Both seeing a colour and seeing an aspect have *duration*.

On the other hand, seeing an aspect is also analogous to interpreting in that seeing an aspect is subject to the will. We can change at will the aspect of a particular object that dawns on us simply by concentrating our attention on certain particulars rather than on others as in the duck-rabbit example.

To this, I would add that, for Wittgenstein, a further similarity existing between seeing an aspect and interpreting is represented by the fact that, in order to see something as, we need to have learned something. We can imagine a baby perceiving a red object, but it does not make sense to say that the baby sees something as a cube.

According to Budd's interpretation, we ought to say that, for Wittgenstein, human reason performs a third type of activity besides imagining and judging: seeing aspects.

If I understand Budd correctly seeing an aspect ought to be, for Wittgenstein, a phenomenon that, having characteristics in common both with seeing and interpreting, ought to be described, in the absence of further qualifications, as a kind of *looking + thinking*. However, this conclusion is explicitly rejected by Wittgenstein:[29]

Is being struck looking plus thinking? No. Many of our concepts *cross* here.

What I take to be one of the clearest accounts given by Wittgenstein of the phenomenon he calls 'dawning of an aspect' is given by the passage quoted at p. 166 from the *Philosophical Investigations*.

From an analysis of the Wittgensteinian text it seems clear to me that, on the one hand, (i) an internal relation is what is at the root of the dawning of an aspect; and that, on the other hand, (ii) the aspect of the object that dawns on us is *not* a property of the object.

Some of the consequences I can derive from (i) and (ii) are that, for Wittgenstein, seeing an aspect is part of the faculty of imagination and, secondly, that since an aspect is not a property of an object, seeing something as is possible, as it is expressed in the quotation from *Philosophical Investigations* at p. 164, because the person who sees something as can do, has learnt, is master of certain techniques, i.e., understands certain concepts.

The correctness of this view is particularly *visible* in the case of mathematical aspects (or patterns) as it results from the following example.

In science we, very often, come across the problem of solving a system of linear equations, for example the system:

---

[29] See [Wittgenstein, 1983], Part II, §xi, p. 211^e.

8. SEEING OR INTERPRETING?

$$\begin{aligned} x_1 - 2x_2 + x_3 &= 1 \\ 2x_1 - x_2 + x_3 &= 2 \\ 4x_1 + x_2 - x_3 &= 1 \end{aligned}$$

The most powerful way of attacking this problem so far found is given by: i) 'seeing the coefficients of the system as a matrix', i.e.,

$$A = \begin{pmatrix} 1 & -2 & 1 \\ 2 & -1 & 1 \\ 4 & 1 & -1 \end{pmatrix}$$

ii) examining the augmented matrix:

$$(A|b) = \begin{pmatrix} 1 & -2 & 1 & | & 1 \\ 2 & -1 & 1 & | & 2 \\ 4 & 1 & -1 & | & 1 \end{pmatrix}$$

and iii) applying concepts and results of matrix theory to find out whether the system of equations has solutions, etc.

If we carefully consider what happens when we are faced with a situation like that described in the example above, we realise that we can *suddenly* see the coefficients of the system of linear equations as a matrix and that such an aspect is not a property of the system of equations, as, instead, would be that individuated by the proposition 'The term on the right hand side of the equality sign of the first equation (from the top) is 1'. (If we change system of representation, we will not see the coefficients of the system of linear equations as a matrix, but 1 will always be at the right hand side of the equality sign of the first equation (from the top) belonging to the system.)[30]

---

[30] The conception, held by Wittgenstein, that certain perceptions are influenced by a learning process or, to put it in a slightly different way, by acquired experience is as old as Helmholtz, who wrote:

> ...to many physiologists and psychologists the connection between the sensation and the conception of the object usually appears to be so rigid and obligatory that they are not disposed to admit that, to a considerable extent at least, it depends on acquired experience, that is, on psychic activity,

(in [Richards, 1977], p. 238); and described the very test used by Wittgenstein to discriminate between *seeing* (*innate sensations*, in Helmholtz's terminology) and *seeing something as* (*sensations which are the product of experience*, in Helmholtz's terminology):

> ...nothing in our sense-perceptions can be recognised as sensation [innate] which can be overcome in the perceptual image and converted into its opposite by factors that are demonstrably due to experience,

On the other hand, it is just as obvious that concepts are here in play as necessary conditions for the aspect/pattern of matrix to become perspicuous when we examine a system of linear equations, that is, at the very least we must know what a matrix is.

If my analysis is correct, what is revealed in Wittgenstein's discussion of aspect seeing is not his belief in some kind of phenomenon which interpolates between imagination and intellect showing the existence of another faculty of reason besides the two already mentioned, but the interesting fact that *concepts reach very deep*.[31]

The correctness of Wittgenstein's position is confirmed by the remarks contained in the following section.

9 Perceptions and expectations

When we truthfully say 'I perceive a', our perceiving a is partly a consequence of our ability to focus on a. On the other hand, such an ability to focus on a depends also on our expectations concerning how a is going to behave if we do such and such; on what a is going to look like if we change our position, etc.

The influence of expectations on focussing and perception has been studied by experimental psychologists. This phenomenon is particularly remarkable in experiments conducted by Johansson in which holistic considerations with regard to the stimuli become crucial to explaining what is perceived. In one of these experiments Johansson:[32]

> ...filmed an actor in the dark with small lights attached to his joints (ankles, knees, hips, shoulders, elbows, and wrists) so that nothing was visible except the lights...
>
> When the actor was seated motionless in a chair, observers perceived a meaningless configuration of points, rather like a constellation of stars. Nobody perceived the lights to be connected to a person. But within fractions of a second after the

---

(see Ibid., p. 240). The main difference between Wittgenstein's position and Helmholtz's seems to me to lie in the conception of the nature of perceptions which, to use Helmholtz's terminology, are not innate sensations. For Helmholtz, perceptions that are not innate sensations are nothing but unconscious conclusions that we draw on the basis of our interaction with the world. These conclusions, and the concepts we form, have a purely empirical basis. For Wittgenstein, instead, there are no hidden psychological mechanisms which generate concepts and unconsciously derive conclusions from a given set of premisses. Concepts, given in terms of their possession conditions, have a public, social dimension which rests on inventing, learning and mastering a number of conventions and techniques, conventions and techniques which come into being at the same time as the concepts they express.

[31] Such a view of Wittgenstein is very similar to the position defended by J. McDowell in [McDowell, 1994] on the same issue, even though in Wittgenstein's treatment of the relationship existing between concepts and perceptions is present no temptation to veer towards idealism.

[32] See [Palmer, 1999], Part III, Ch. 10, §4.1. p. 511.

9. PERCEPTIONS AND EXPECTATIONS

actor began to move, first standing up and then walking, he was immediately and unmistakably perceived as a person in motion.

If it is now clear that Wittgenstein's remarks on aspect seeing are not just philosophical sophistry, but that there exists a well documented object of psychological investigation connected with them which goes back in time to Köhler and the *Gestalt* psychologists, we must ask ourselves what does aspect seeing show about the rôle of concepts in perception. Do we have to consider concepts as supplementary senses, as what produces magnifications and extensions of the senses, or what exactly? Perhaps, a consideration of the example of seeing a mathematical pattern given in §5.8 can help us to identify the function of concepts in perception.

The example of section 5.8 presents us with the purest form of organisational function that concepts perform in structuring representation. If we observe the system of equations, the *seeing* part of the act of 'seeing the coefficients of the equations belonging to the system as a matrix', has no relevance whatsoever, in the sense that if we knew no mathematics, not only we would be unable to see the coefficients of the equations as a matrix, but, as it happens when we become acquainted for the first time with scripts which belong to languages of which we even ignore the shape of the symbols, we could very easily see the equations themselves as decorative motifs or as the outcome of idle doodling.

The fact that we can safely generalise this interpretation to the *seeing* of any mathematical pattern leads us to conclude that, although aspect seeing is a phenomenon characteristic of perception, it does not belong to the sensory part of it. It is what brings purely sensory input into a manifold and the extremely interesting feature of this (in Kantian sense) is that concepts perform pre-reflectively (without being involved in judgments) the structuring/organisational rôle which brings sensory input into a manifold.

According to this interpretation, concepts, besides presiding to the judgement making activity of the intellect, perform a function very similar to that of the Kantian *a priori* pure intuitions of space and time. They provide structure (organisation) to the sensory input.

The model of experience which derives from these results is very different from that inherited from the post-Kantian tradition. I will list below some of its most remarkable characteristics.

First, although the perceptual and intellectual faculties of reason remain distinct and there is no justification for presupposing the existence of a third faculty of reason interpolating between these two, some of the theoretical vehicles through which such faculties are exercised (concepts) are by them shared.

Secondly, such shared theoretical vehicles are not given *a priori* in the mind, etc., but they are rather the outcome of the cultural activity of the human kind. This is a very important point, because it shows that perceptions are influenced by external, non-psychological factors. What this implies is, among other things, that not only language has a very important social aspect, but so does also the mind.

One of the philosophical consequences of the phenomenon I have called 'seeing something as' is that, it generates a very plausible epistemology of mathematics as a science of patterns.

In fact, if mathematical patterns are, as I have argued, the structure (form) of our perception of a concrete object (or of an aspect of a concrete object, etc.), there is a very plausible explanation of the way to access such patterns: we simply 'have them', that is, they are given to our consciousness in our perception of 'something as ...'

Therefore, the view according to which mathematics is a science of patterns, on the one hand, is a form of realism about mathematics which shows no Platonist inclinations — patterns are neither objects nor properties of objects — and, on the other hand, smoothly harmonizes with an which naturally arises from it.

10 Psychologism?

At this point someone may begin to wonder how a view of mathematics such as the one defended in this chapter could avoid psychologism.

Indeed, now a Fregean philosopher might stick his neck out and warn us against the dangers inherent in veering towards psychologism quoting Frege, who famously said:[33]

> No, sensations are absolutely no concern of arithmetic. No more are mental pictures, formed from the amalgamated traces of earlier sense-impressions. All these phases of consciousness are characteristically fluctuating and indefinite, in strong contrast to the definiteness and fixity of the concepts and objects of mathematics. It may, of course, serve some purpose to investigate the ideas and changes of ideas which occur during the course of mathematical thinking; but psychology should not imagine that it can contribute anything whatever to the foundation of arithmetic.

To this it is possible to reply that, first, since mathematics is part of our culture and is something we learn and teach to others on the basis of a number of psycho-socio-cultural pre-conditions, paying attention to what happens when we actually learn such-and-such and to the psycho-socio-cultural pre-conditions of this learning might be a safeguard against the construction of a mythology about numbers, etc. and might also provide a solid base on which to build a plausible epistemology for mathematics.

---

[33][Frege, 1884], Introduction, pp. v-vi.

10. PSYCHOLOGISM?

Secondly, since the 'seeing of an aspect' is something given through a system of representation, and a system of representation is a structured set of concepts and beliefs (notions), it follows that a system of representation cannot be conceived as what makes our senses sharper or more focused, but as one of the factors at the root of the intersubjectivity of mathematical patterns.

In fact, the meaning of notions can be objectively established through a specification of their condition of use and, moreover, notions are not magnifying lenses, but moulds in which objects fit or do not fit. The metaphor of the mould becomes unsatisfactory, if we ignore that the *direction of fit* is from the mould to the object and not *vice versa*. What I mean by this is that it does not make any sense to ask whether something can be seen as a circle or not independently of the existence of a geometrical system of representation.

This is an important point to put to rest any worry concerning the danger of undermining mathematical objectivity through the use of psychological notions in the philosophy of mathematics. For, very much like the kantian pure intuitions of space and time, the presence of a mathematical system of representation provides a pre-reflective structuring of perception common to all those who have adopted the system. Such a position is also able to get rid of the profoundly unsatisfactory naïve empiricist account of the process of concept formation providing at the same time a compelling explanation of what is commonly known as mathematical intuition.

An interesting corollary of the point just made, is that the pre-reflective representation which is at the heart of the dawning of an aspect cannot be accounted for in terms of an Husserlian intuition of an essence.

In fact, independently of the observation that patterns are not conceivable as essences in Husserl's sense, we have seen, in §3.8, that, for Husserl, the intuition of an essence is a process of abstraction, based on what Husserl calls 'phenomenological reduction', whereby the essence is actually given to us 'as an object we can look at'.

But, since, as we mentioned above, 'it does not make any sense to ask whether something can be seen as a circle or not independently of the existence of a geometrical system of representation', it follows that the dawning of an aspect cannot be the outcome of a process of abstraction, and that, consequently, it cannot be accounted for in terms of an Husserlian intuition of an essence.

Lastly, if mathematics, as I have argued, is a science of patterns, it does not follow that we must then be captive of a form of psychologism about mathematics. For, strictly speaking, mathematical activity consists in proving theorems about those structures, which become perspicuous to us when we have a mathematical theory as a system of representation, *regardless the way these are given to us.*

In particular, an explanation of the way mathematical structures are given to us has nothing to do with the grounds for the justification of mathematical assertions. Mathematical assertions are, in fact, justified exclusively on the basis of *a priori* grounds.

The rôle of aspect seeing in mathematical activity, consequently, belongs to the so-called 'context of discovery'. The dawning of mathematical aspects is, as I have argued, a cognitive process in which concepts have a very important function. The mathematician imposes a system of representation on perceptual input and then sees certain things as . . . Now the ability to see certain things as . . . , which follows on the adoption of a system of representation, is an invaluable means to explore the existence and properties of structures which appear to be realized as the forms of our representations. But, of course, in doing so we simply discover (not prove) how concepts are *a priori* related to one another.

11 Two objections

There are a number of objections to the idea of applying an epistemology based on the phenomenon of seeing an aspect to the mathematical case. The first of those I am going to consider is the following. When a mathematical aspect dawns on us, this is not a genuine case of aspect seeing. In fact, it is not sufficient to shift the focus of our attention from one part of the object perceived to another for a mathematical aspect to dawn on us.

A quick reply to the objection above is that in clear-cut cases of aspect seeing like in the duck/rabbit, young woman/old woman, etc. examples, it is not sufficient to shift the focus of our attention from one part of the object perceived to another for the duck (rabbit)-aspect, young (old) woman-aspect, etc. to dawn on us either. For also in these cases, to change the aspect perceived by shifting the attention from one part of the object perceived to another, we must know what a duck, a rabbit, an old woman, and a young woman are.

However, there is a better reply than the one I have just given. And this goes as follows. In contrast with what asserted in the objection above, there are in mathematics aspect seeing phenomena which are relevantly similar to the duck/rabbit, young woman/old woman *Gestalt*-shifts. For example, given the system of equations of page 181, if we shift our attention from seeing the equations as wholes to focussing on the coefficients of their terms, and we know what a matrix is, we will see such coefficients as forming a matrix. And, if we switch back to focussing on the equations as wholes, the matrix of coefficients will then become 'invisible' again.

11. TWO OBJECTIONS

A similar example is given by the cube in fig. 5.2.

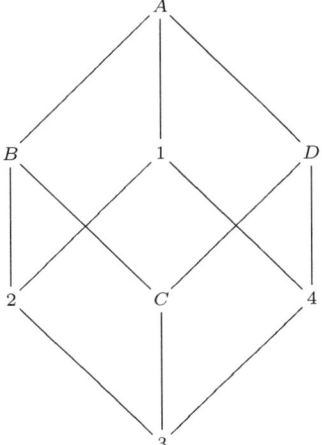

Figure 5.2.

Here, simply by shifting the focus of our attention from the vertex 1 to the vertex C, we can see the face $ABCD$ 'coming forward' in one case, and the face 1234 in the other.

With regard to the object drawn in fig. 5.2, we have an even more interesting *Gestalt*-shift than the one given above. In fact, if we change system of representation from 3-dimensional Euclidean geometry to lattice theory, we can see the object in fig. 5.2 as a lattice. This would be a genuine case of *Gestalt*-shift, because to a change in the system of representation there does not correspond a physical change in the object represented.

Indeed, a change in the system of representation simply helps us to focus on what we regard as parts of the object and on relations defined between them. For example, within 3-dimensional Euclidean geometry, we could describe the object in fig. 5.2 in relation to its 'faces' (F), 'angles' (A), 'edges' (E), 'vertexes' (V), etc. and we could say that it is a special type of polyhedron called 'cube' which, among others, has the property that:

$$V - E + F = 2.$$

Whereas within lattice theory we could describe the object in fig. 5.2 as a distributive lattice, and in the theory of Boolean algebras as a Boolean algebra.

The second objection against the applicability to mathematics of an epistemology based on the phenomenon of seeing an aspect is that seeing an aspect always presupposes *seeing*, i.e., perception, but there are many cases in which the objects of mathematical investigation cannot even be drawn. Take the example of a function f, for $f : [a, b] \mapsto \mathbb{R}$ and $[a, b] \subseteq \mathbb{R}$, which is continuous at every point, but is not differentiable at any point; or that of a function g, where $g : [a, b] \mapsto \mathbb{R}$ and $[a, b] \subseteq \mathbb{R}$, which is such that for any $x \in [a, b]$,

$$g(x) = \begin{cases} 0 & \text{if } x \text{ is rational,} \\ 1 & \text{otherwise.} \end{cases}$$

If these are mathematical patterns which cannot be drawn, that is, which cannot be given within perceptual space, so the objection goes, how can these patterns dawn on us?

The answer to this question is that there are many examples in mathematics of aspects of objects, which dawn on us when a non-perceptual representation of these objects is given. For instance, if we assume that the representation of a curve in $\mathbb{R}^2$ by means of an equation is non-perceptual then, given some lines in $\mathbb{R}^2$ described by a system of equations, through algebraic manipulations of such a system, it might dawn on us that the lines must intersect. But, of course, the dawning of an aspect, or equivalently, having an intuition that . . . , is not sufficient to justify our assertion that the lines must intersect and, therefore, at this point we are required to provide a proof that this is what is shown by our algebraic manipulations.

The great advantage of producing an algebraic representation of geometry is the well known phenomenon whereby we replace the extremely difficult task of producing a representation of certain geometrical problems within perceptual space with the comparatively easier one of surveying a clear and compact algebraic translation of these problems.

A particularly interesting example which illustrates this point is one of those I mentioned above, and, precisely, that which concerns functions of a real variable which are everywhere continuous and nowhere differentiable.

If we consider the function $f : \mathbb{R} \mapsto \mathbb{R}$, such that $f(x) = \{x\}$, where $\{x\}$ refers to the distance from x to the nearest integer, we can define the sequence of functions (f_n), where $f_n : \mathbb{R} \mapsto \mathbb{R}$ and $n \in \mathbb{N}$, such that

$$f_n(x) = \frac{1}{10^n}\{10^n x\}.$$

Notice that the function f and each function f_n are continuous functions the graphs of which are easy to draw or imagine. From their graphs we can see that these functions are continuous everywhere and non-differentiable at infinitely many points.

11. TWO OBJECTIONS

However, if we consider the function $\mathcal{F}$, such that $\mathcal{F}: \mathbb{R} \mapsto \mathbb{R}$ and

$$\mathcal{F}(x) = \sum_{n=1}^{\infty} \frac{1}{10^n} \{10^n x\},$$

a moment's reflection will make us realize that we cannot draw its graph, because, although $\mathcal{F}$ is everywhere continuous, it is nowhere differentiable. But, if we cannot draw the graph of $\mathcal{F}$, how does it dawn on us that $\mathcal{F}$ is continuous everywhere and nowhere differentiable?

Well, the first step in the process must be the observation that the points at which the partial sum $f_1 + \cdots + f_n$ is not differentiable increase monotonically with the increase of n without any compromission of the continuity of $f_1 + \cdots + f_n$.

The second step is seeing that the sum of the series $\sum_{n=1}^{\infty} f_n$ must be a function which is everywhere continuous and nowhere differentiable.

It is important to realize that the only tools used in the proof that the conjecture we formulate about $\mathcal{F}$ is correct are 'arithmetical' in the sense that it is possible to understand the proof without appealing to geometrical representations which can be carried out within perceptual space.

Indeed, it is the definition of f_n what grants that the sequence of functions (f_n) passes the Weierstrass M-test,[34] and that, consequently, $\sum_{n=1}^{\infty} f_n$ converges uniformly to $\mathcal{F}$

$$\mathcal{F}(x) = \sum_{n=1}^{\infty} f_n(x) = \sum_{n=1}^{\infty} \frac{1}{10^n} \{10^n x\}.$$

Now, since $\sum_{n=1}^{\infty} f_n$ converges uniformly to $\mathcal{F}$, and each f_n is everywhere continuous, it follows that $\mathcal{F}$ is everywhere continuous. The proof that $\mathcal{F}$ is nowhere differentiable is also purely 'arithmetical'.

From the considerations above follows that the ability to represent such functions as $\mathcal{F}$ and the possibility of studying them providing, among other things, proofs of their existence is a feat which can only be accomplished within the language of analysis. Of course, to be able to represent such counterintuitive entities we have to have the concepts of function, continuity, sequence, series, convergence, differentiability, etc. But, it is the language of analysis the happy medium through which mathematicians give precise expression to such concepts making in this way possible the objective study of the relations in which they are with one another, the formulation of new key concepts like that of uniform convergence, etc. In other words, it is the language of analysis what enables us to overcome the limitations

---
[34] See on this [Spivak, 1992], ch. 23, p. 420.

of perceptual representation without compromising the ability to produce pre-reflective representations. And this is possible, because the language of analysis, besides performing, as Brouwer thought, the rôle of a means for communicating and storing mathematical results, is a conceptual notation in Frege's sense, that is, the language of analysis is a system of representation for the concepts of analysis and their relations.

I must here stress that, even in a situation in which we are dealing with non-perceptual representations of objects, it makes sense to talk about the dawning of an aspect, that is, it makes sense to talk about a phenomenon which entails, ultimately, an involvement of our perceptions, because for the aspect of a particular object (or for the aspect of an aspect of ...) to dawn on us through manipulations of the formalism related to it, we need to perceive the symbols contained in the formalism (or imagine them). And, in actual fact, what dawns on us is an aspect of the formalism, which has great importance, because it corresponds to an aspect of the entities we do not perceive.

12 Sets, patterns, and infinity

Taking for granted the correctness of the account given of the rôle of concepts in the formation of mathematical patterns, it makes sense to ask how this works in particular with sets. For in this book I am defending a form of set-theoretical realism.

Since:[35]

> A pack of wolves, a bunch of grapes, or a flock of pigeons are all examples of sets of things

it is easy to see how sets are entities of which we have experience.

As Ullman says in what follows, sets (classes) perform a fundamental rôle even in visual recognition:[36]

> It seems to me that, for the purpose of visual recognition, the natural place to start are classes of objects in the world that exist as "relative classes" rather than "absolute classes". This means that rather than some absolute, objective classes, the formation of object classes depends on a selected set of attributes of interest. For a biological system, attributes of interest may depend on distinctions such as whether the objects are edible or not, whether or not they can move on their own, and how fast, whether they are graspable, and so on. Given a set of attributes, a measure of similarity between objects can be defined, and this can lead to the formation of natural clusters of objects that share similar attributes ... This clustering is defined in the world, not in the image, since the attributes of interest are not necessarily directly measurable in the image. The clustering depends, however, on the choice of relevant attributes; different attributes will give rise to different clustering.

---

[35] See [Halmos, 1974], Section 1, p. 1.
[36] See [Ullman, 1997], Ch. 6, pp. 187–188.

12. SETS, PATTERNS, AND INFINITY

But, of course, the experience we have when we perceive certain things 'as a set of ...' is one of aspect seeing, because our perception of a set is not reducible to what we might call 'raw sensory input'. Indeed, our perception of a set has a form which, besides being dependent on the object(s) described, depends also on a system of representation, that is, on the system of beliefs to which what Ullman calls 'attributes of interest' belong.

If, for example, we had two systems of representation which included the concepts: flock of sheep, white sheep, and speckled sheep; but within each of which the above mentioned concepts had a different *weighting*, we could then see the same sensory input in one case as one flock of sheep, and in the other as two flocks of sheep — the flock of white sheep and the flock of speckled sheep — switching at will from one aspect to the other in relation to how important the concepts of 'white' and 'speckled' are with regard to the concept of 'flock' in the system of representation adopted.

However, taking for granted that a good case can be made in favour of the idea that finite sets are strongly related to part of our experience, also in the sense of being directly involved in the process of visual recognition, what remains to be explained is the relation, if any, between transfinite sets and experience.

Indeed, there seems to be a difficulty here which is related to the third, and last, objection I am going to consider against the applicability to mathematics of an epistemology based on the phenomenon of seeing an aspect. The objection goes as follows. If what I call 'mathematical patterns' are aspects of concrete objects, to be able to say that there are, for instance, infinitely many natural numbers, I would need to know either that there are infinitely many physical objects or that infinity is a feature of concrete objects.[37] But, as Hilbert has pointed out, in the age of the theory of relativity and of quantum mechanics both these traditional empiricist routes to the concept of infinity have become impracticable and this not as a consequence of metaphysical speculation, but as an outcome of the investigations of the empirical sciences.[38]

---

[37] Indeed, for any empirical or quasi-empirical view of mathematical knowledge, the *very large finite* is, besides the infinite, a potential source of problems. How do we gain information about natural numbers having, for instance, a size greater or equal to $10^{1000^{1000}}$? or of polygons with a thousand sides? or of geometrical figures describable only in n-dimensional metric spaces, for $n > 4$? etc.

However, since, if we are able to solve the problem of providing an account of our knowledge of infinite magnitudes, this will imply a similar solution for our knowledge of the *very large finite*, I will not tackle the latter problem in this book.

[38] [Hilbert, 1926], pp. 185–186:

One's first naïve impression of natural events and of matter is one of permanency, of continuity. When we consider a piece of metal or a volume of liquid, we get the impression that they are unlimitedly divisible, that their smallest parts exhibit the

The solution to the problem concerning the possibility of giving a plausible account of the use of the concept of infinity in mathematics in general, and in set theory in particular, within the view of mathematics defended in this book, originates from the consideration that some patterns are 'patterns

---

same properties that the whole does. But wherever the methods of investigating the physics of matter have been sufficiently refined, scientists have met divisibility boundaries which do not result from the shortcomings of their efforts but from the very nature of things. Consequently we could even interpret the tendency of modern science as emancipation from the infinitely small. Instead of the old principle *natura non facit saltus*, we might even assert the opposite, viz., "nature makes jumps".

It is common knowledge that all matter is composed of tiny building blocks called "atoms", the combinations and connections of which produce all the variety of macroscopic objects. Still physics did not stop at the atomism of matter. At the end of the last century there appeared the atomism of electricity which seems much more bizarre at first sight. Electricity, which until then had been thought of as a fluid and was considered the model of a continuously active agent, was then shown to be built up of positive and negative *electrons*.

In addition to matter and electricity, there is one other entity in physics for which the law of conservation holds, viz., energy. But it has been established that even energy does not unconditionally admit of infinite divisibility. Planck has discovered *quanta of energy*.

Hence, a homogeneous continuum which admits of the sort of divisibility needed to realize the infinitely small is nowhere to be found in reality. The infinite divisibility of a continuum is an operation which exists only in thought. It is merely an idea which is in fact impugned by the results of our observations of nature and of our physical and chemical experiments.

The second place where we encounter the question of whether the infinite is found in nature is in the consideration of the universe as a whole. Here we must consider the expanse of the universe to determine whether it embraces anything infinitely large. But here again modern science, in particular astronomy, has reopened the question and is endeavoring to solve it, not by the defective means of metaphysical speculation, but by reasons which are based on experiment and on the application of the laws of nature. Here, too, serious objections against infinity have been found. *Euclidean* geometry necessarily leads to the postulate that space is infinite. Although Euclidean geometry is indeed a consistent conceptual system, it does not thereby follow that Euclidean geometry actually holds in reality. Whether or not real space is euclidean can be determined only through observation and experiment. The attempt to prove the infinity of space by pure speculation contains gross errors. From the fact that outside a certain portion of space there is always more space, it follows only that space is unbounded, not that it is infinite. Unboundedness and finiteness are compatible. In so-called *elliptical* geometry, mathematical investigation furnishes the natural model of a finite universe. Today the abandonment of Euclidean geometry is no longer merely a mathematical or philosophical speculation but is suggested by considerations which originally had nothing to do with the question of the finiteness of the universe. Einstein has shown that Euclidean geometry must be abandoned. On the basis of his gravitational theory, he deals with cosmological questions and shows that a finite universe is possible. Moreover, all the results of astronomy are perfectly compatible with the postulate that the universe is elliptical.

We have established that the universe is finite in two respects, i.e., as regards the infinitely small and the infinitely large.

12. SETS, PATTERNS, AND INFINITY

of patterns'.

If I say 'There are three chairs', what I express in my use of the term 'three' captures a numerical pattern of a set of concrete objects, and although the pattern given by 'three' is neither an object nor is a property of an object, it is objectively given in the sense that I can *prove* that there are three chairs in the room.

The fact that a pattern such as 'three' is objectively given is what provides the conditions of identification for it, i.e., it is what provides the necessary and sufficient conditions for referring to the pattern denoted by 'three' as a member of another pattern.

Therefore, we have that whereas numerical patterns such as: one, two, three, etc., may be considered as aspects of sets of concrete objects, or as the outcome of operations performed on such ones; the introduction of the concept of (actual) infinity comes about as a consequence of being able to see the collection of *all* those numerical aspects as a whole.

It is important to notice that the phenomenon described here as 'seeing a collection as a whole' is an instance of aspect seeing in which we are dealing with a pattern of patterns. The correctness of this view is confirmed by the consideration that the ability to see the collection of (the set-theoretical representations of) the natural numbers as a whole (complete totality or set) depends on the system of representation adopted.

In fact, it is only through a modification of the Euclidean concept of 'whole' — the whole is greater than the part (Euclid's axiom 5) — and the axiomatization of set theory which eliminates the known paradoxes, that it becomes mathematically meaningful to see infinite collections not as entities of unbounded growth, but as infinite totalities, i.e., as entities such that, given any two of them, it makes sense to ask whether the number of elements of one is lesser, equal, or greater than the number of elements of the other.

As a confirmation of this we have the exemplary case of Galileo. Galileo, found himself in an embarrassing situation when he observed that, if we consider N as a completed totality, there appear to be as many squares of natural numbers as there are natural numbers. This is the case, he argued, because each natural number is the root of a square number and there are as many squares of natural numbers as their roots.

The observation above was a source of embarrassment for Galileo, because, besides contradicting Euclid's axiom 5, it seemed to conflict with the other observation that the occurrence of perfect squares in the natural ordering of N becomes more and more rare the larger the initial segment of N we consider.[39] For, the latter observation suggested to him that most

---

[39] In the first 100 natural numbers there are only 10 squares, in the first 10000 only

natural numbers are not squares of natural numbers.[40]

At this point Galileo, unable to deny Euclid's Axiom 5 and unwilling to consider number theory to be inconsistent, to save the day, decided to reject the notion of actual infinity in favour of the old Aristotelian concept of potential infinity, and adopt an *ad hoc*, monster barring, deliberation whereby[41]

> **Salv.** ... attributes of equal greater and lesser are not applicable to infinities, but only to finished [terminating, completed] quantities.

100, in the first 1000000 only 1000, etc.

[40][Galileo, 1638], Giornata Prima, pp. 78–79:

Salv. Ma se io domanderò, quante siano le radici, non si può negare che elle non siano quante tutti i numeri, poichè non vi è numero alcuno che non sia radice di qualche quadrato; e stante questo, converrà dire che i numeri quadrati siano quanti tutti i numeri, poichè tanti sono quante le lor radici, e radici son tutti i numeri: e pur da principio dicemmo, tutti i numeri esser assai più che tutti i quadrati, essendo la maggior parte non quadrati. E pur tuttavia si va la moltitudine de i quadrati sempre con maggior proporzione diminuendo, quanto a maggior numeri si trapassa; perchè sino a cento vi sono dieci quadrati, che è quanto a dire la decima parte esser quadrati; in dieci mila solo la centesima parte son quadrati, in un milione solo la millesima: e pur nel numero infinito, se concepir lo potessimo, bisognerebbe dire, tanti essere i quadrati quanti tutti i numeri insieme.

[41]See [Galileo, 1638], Giornata Prima, p. 79.

CHAPTER 6

MATHEMATICS: A QUASI-EMPIRICAL SCIENCE?

1 Introduction

In Chapter 5 it was argued that mathematics is a science of patterns and that such a view of mathematics is able to answer satisfactorily both the questions 'What is the nature of mathematical reality?' and 'How do we acquire mathematical knowledge?'.

Since, in accepting the conclusions of Chapter 5, I am committed to the idea that mathematics is a science, the problem that now has to take centre stage is that of determining whether or not mathematics is conceivable as an empirical science.

To show that this is an important, live issue within contemporary philosophy of mathematics, I will defend the idea, already put forward in [Kalmár, 1967], that the failure of the traditional programmes in the foundations of mathematics justifies the view according to which all the most serious attempts made between the end of the XIXth century and the end of the XXth century to prove that mathematical knowledge is non-empirical have come to nothing.

I shall then analyse Mill's, Kitcher's, and Gillies's arguments in favour of the idea that mathematics is an empirical science, and provide reasons why they all fall short of the target.

At this point in the discussion, I shall argue that mathematics is a quasi-empirical science in the sense that, although mathematical statements are *a priori* true or false, in mathematics, as in the empirical sciences, it makes sense to say that mathematical statements and theories can be refuted.

The correctness of the view that mathematical statements and theories can be refuted will then be confirmed by a discussion of a case-history in which it will be shown that, there are times in the historical development of Euclidean geometry when mathematical knowledge grows in a non-cumulative way. Indeed, evidence will be provided in favour of the idea that, when we look at the historical development of Euclidean geometry, we realize that this branch of mathematics consists of a finite sequence of refutable theories.

The chapter will end with a discussion of the notion of Lakatosian Mathematical Research Programme and with a section containing considerations concerning whether a quasi-empiricist view of mathematical knowledge is compatible with research in the foundations of mathematics and with the view of mathematical justification known as 'foundationalism'.

2 A basic distinction

David Hume in discussing what human understanding applies to distinguishes between *matters of fact* and *relations of ideas*. He says that:[1]

> The understanding exerts itself after two different ways, as it judges from demonstration or probability; as it regards the abstract relations of our ideas, or those relations of objects, *of which experience only gives us information* [matters of fact].

Kant takes on board the Humean distinction. In fact, he explains the difference between mathematics and, for instance, physics by saying that mathematical statements are *a priori* true (or false), whereas statements belonging to physics are *a posteriori* true (or false).

With Frege the distinction is no longer drawn between statements belonging to mathematics and statements belonging to the empirical sciences, but between statements belonging to logic/arithmetic and those belonging to the other (empirical) sciences.

What emerges from the views of Hume, Kant, and Frege is therefore that, in talking about the sciences, it makes sense to distinguish the sciences which produce knowledge of what only experience can give us information about — the so-called 'empirical sciences' — from those which produce knowledge of what experience is not the only source of information.

3 Anti-empiricism in mathematics

One of the consequences of the distinction between empirical and *a priori* sciences described in the section above was that a large part of the philosophical community thought mathematics to be an *a priori* science. And, consequently, as Kalmár says,[2]

> ...research in the Foundations of Mathematics [was] pursued with the supposition that mathematics is a pure deductive science, and with the hope that we shall show it to be a *firmly founded* pure deductive science.

This is precisely the attitude of the logicists, of the followers of Hilbert's programme, and of the constructivists.

---
[1][Hume, 1739], Book II, Part III, §III, pp. 413-414.
[2][Kalmár, 1967], §7, in: [Lakatos, 1967], pp. 193.

4. IS MATHEMATICS AN EMPIRICAL SCIENCE?

In fact, for Frege arithmetic is part of logic, and arithmetical statements are analytic in the sense that their proofs involve only laws of logic and definitions of a logical nature. This, of course, implies that arithmetical statements are *a priori* true (or false), because their proofs (or refutations) do not appeal in any way to matters of fact, and that consequently, Logicism is incompatible with empiricism.

Moreover since, according to a follower of Hilbert's programme, the only thing that matters in proving the consistency of a mathematical theory T is the logical form of T, and such a logical form becomes transparent by means of the formalization of T, it follows that any consideration relating to the origin and/or meaning of mathematical statements belonging to T is seen by an Hilbertian as irrelevant or even as potentially dangerous. For it obscures the only thing that matters to the articulation of proof theory: the logical form of T. Consequently, empiricism turns out to be incompatible also with Hilbert's programme.

Lastly, although constructivism is correctly describable as a form of mathematical verificationism, it is important to notice that constructivism is, nevertheless, incompatible with empiricism. The reason for this is that for the constructivist mathematics is the outcome of a creative act, whereas for the empiricist mathematics is related to our experience of reality.

But now, since, as a consequence of the failure of Logicism, Hilbert's programme, and Intuitionism, the hope to show that mathematics is a firmly founded pure deductive science has not been realized, should we then conclude with Kalmár that[3]

> ... mathematics, like other sciences, is ultimately based upon, and has been tested in, practice?

And what does this mean anyway?

4 Is mathematics an empirical science?

Let us assume, for the sake of argument, that the assertion 'mathematics is ultimately based upon, and has been tested in, practice' means that mathematics is an empirical science, and discuss whether this position is tenable.

Although, as we have already seen, for Hume and Kant mathematics in general, and for Frege arithmetic only, are not conceivable as empirical sciences, there have been authors for whom this is not the case. The classical expression of the view which considers mathematics to be an empirical science is J. S. Mill's.

For Mill:[4]

---

[3][Kalmár, 1967], §7, in: [Lakatos, 1967], pp. 192.
[4][Resnik, 1980], Ch. Four, p. 137.

> ...mathematics is known inductively from experience. Mathematical reasoning — indeed, all so-called deductive reasoning — is in reality inductive. The general axioms of mathematics are inductions derived from experience, and its existential postulates assert matters of physical fact. Moreover, there are no special mathematical objects, for mathematics is simply a very general theory of ordinary physical objects.

One of the consequences of Mill's position is that numbers are '... physically perceptible properties of collections of physical objects ...'[5]

Such a view of numbers has been successfully attacked by Frege, who argued that we can see a collection of boots as 'two boots' or as 'a pair of boots', but if the number to be associated with the collection of boots were a physical property of the collection like colour, we would have no choice concening how to describe numerically the collection.[6]

Although Frege's objection is able to show the inadequacy of Mill's position on natural numbers, there are other possible empiricist accounts of numbers which are immune to Frege's criticism. For instance, for Kitcher, who sees arithmetic as based on the operations of segregation and combination of objects, we have that because there are different ways of combining and segregating objects, we can in one case consider a collection of two boots as a pair of boots, and in the other as two boots.[7]

Moreover, for Gillies, since: (1) we perceive sets; (2) there are different ways of perceiving sets; and (3) numbers are properties of sets; we can then perceive a collection of two objects in one case as a set of two boots and in the other as two sets of one boot each.[8]

Another powerful objection made by Frege to Mill's point mentioned above goes as follows: if numbers are properties of collections of physical objects how do we explain the origin of very large numbers?

Once again the objection is unanswerable by Mill, because, given our bio-technological limitations, at any one time there exists an upper bound concerning the maximum size of the finite set of physical objects that we can experience, but there is no upper bound concerning the natural numbers we can do arithmetic with.

However, such an objection can be answered by other empiricist accounts of mathematics. For instance, for Kitcher, since collecting is an operation which is not limited to objects, but can be applied to what has already been collected to form higher-order collections, we can then, in principle, generate arbitrarily large collections.[9]

---

[5][Resnik, 1980], Ch. Four, p. 155.
[6][Frege, 1884], §25, pp. 32–33.
[7]See [Kitcher, 1984], Ch. 6, p. 108.
[8]See [Gillies, 2000].
[9]See [Kitcher, 1984], p. 129.

4. IS MATHEMATICS AN EMPIRICAL SCIENCE?

Gillies, on the other hand, is of the opinion that we are justified in believing in the existence of arbitrarily large numbers, because such a belief is confirmed by the use made of such numbers in the empirical sciences.

The last criticism made by Frege of Mill'sphilosophy of mathematics that I am going to examine in this section is that, in contrast with what Mill asserts, arithmetical laws cannot be conceived as inductive generalizations. Frege argues that:[10]

> Induction (...properly understood) must base itself on the theory of probability, since it can never render a proposition more than probable. But how probability theory could possibly be developed without presupposing arithmetical laws is beyond comprehension.

Indeed, if we consider the very strong link existing between induction and probability, there does not seem to be a satisfactory reply to this objection which can come from Mill's position. The only way of avoiding the objection seems to be that of modifying the empiricist standpoint so that it does not imply that the laws of arithmetic are inductive generalizations. This is, in fact, the strategy followed by Kitcher, for whom:[11]

> ...mathematical truths are true in virtue of stipulations which we set down, specifying conditions on the extensions of predicates *which actually are satisfied by nothing at all but are approximately satisfied by operations we perform (including physical operations)*.

On the other hand, according to Gillies:[12]

> We can still agree with Mill that simple equations, involving small numbers, such as $2+3=5$ were obtained first as empirical generalizations of the results of manipulating pebbles, etc.

but, when we consider arithmetic as a whole, we realize that this no longer holds in general, and that arithmetic consists of statements which, although not empirically falsifiable, can be confirmed (disconfirmed) by experience. Consequently, for Gillies, the belief in arithmetical statements 'is justified not by *a priori* proofs, but by the fact that they are well-confirmed by experience'.[13]

However, there is a *prima facie* obstacle in the way of a possible generalization of Gillies's view to the whole of mathematics. And such an obstacle is represented by all those mathematical statements — great many of which belong to transfinite set theory — which, not being indispensable to our best scientific theories, are neither confirmable nor disconfirmable by experience.

---

[10] [Frege, 1884], §10, pp. 16–17.
[11] See [Kitcher, 1984], p. 110.
[12] See [Gillies, 2000], p. 44.
[13] See [Gillies, 2000], p. 45.

Gillies has a way of coping with such a situation within his philosophy of mathematics. In fact, he applies the well known distinction between science and metaphysics to mathematics. For him a theory T is scientific if its statements are confirmable (disconfirmable) by experience, otherwise T is metaphysical.

Applying the scientific/metaphysical distinction to mathematical theories, Gillies can free his empiricist philosophy of mathematics from potential counterexamples relegating all those mathematical statements which are neither confirmable nor disconfirmable by experience to the domain of metaphysics. This is a domain the elements of which have content, but are neither true nor false, and, therefore, do not produce knowledge.

Although Frege's objections to Mill's prove to be unanswerable within Mill's system, they are far from showing that any empiricist philosophy of mathematics is untenable. As we have seen above, both Kitcher and Gillies have formulated empiricist standpoints in the philosophy of mathematics which are resistant to Frege's critical acid.

However there is an obstacle which neither Kitcher nor Gillies can overcome, and this is what I call the 'empiricist's dilemma'.

5 The empiricist's dilemma

For someone who believes that mathematics is an empirical science, mathematical knowledge must be ultimately based on experience in the sense that experience is the ultimate and indispensable ground of justification for the assertion of a mathematical statement.

In contrast with the above empiricist constraint on knowledge, the mathematical community accepts a mathematical statement to be part of a mathematical theory not because this has been empirically confirmed — through experimental tests, or the indispensability of such a statement to a successful theory belonging to an empirical science, or a probabilistic argument, etc. — but on the basis of a mathematical proof.

A relatively recent and well known example of such acceptability conditions at work is that of what we can now call 'Fermat's Last Theorem', that is, the statement according to which, if x, y, z, n are positive integers and $2 < n$, the equation

$$x^n + y^n = z^n$$

has no solutions.

Indeed, mathematicians had plenty of confirmations of the correctness of this assertion for about the last 300 years, but it was only when Wiles produced his proof that the statement above was, eventually, accepted as a theorem of number theory.

5. THE EMPIRICIST'S DILEMMA

From these considerations follows that, if we are empiricists and consider compelling the criteria for the acceptability of mathematical statements adopted by the mathematical community, we must conclude that either there is no such a thing as mathematical knowledge, or that there is mathematical knowledge, but we cannot give an empiricist account of it. This is the empiricist's dilemma.

It is clear that neither of the two alternatives present in the above dilemma is acceptable to the empiricist. For, the second implies giving up on empiricism as the correct and general account of knowledge; whereas the first leads to a clear conflict with a commonly held view about the significance of the indispensability of mathematics to the empirical sciences. According to such a view, the indispensability of a mathematical theory to successful theories belonging to the empirical sciences, commits one to the belief in the existence of the entities postulated by the mathematical theory, and, consequently, commits one also to the belief that the mathematical theory in question produces knowledge (see §2.9).

But, at this point someone might ask 'If the indispensability of a mathematical theory to a successful theory belonging to an empirical science "is evidence in favour of the idea that the mathematical theory in question produces knowledge", doesn't this imply that experience is the ultimate and indispensable ground of justification for the assertion of a mathematical statement?'.

The question above must be answered with a 'No'. For, although the indispensability of a mathematical theory to a successful theory belonging to a given empirical science is evidence for the existence of mathematical knowledge, it says nothing about the nature of such a knowledge.

Now, since both Kitcher and Gillies believe in the existence of mathematical knowledge and their views of mathematics have as a consequence that the belief in mathematical statements[14]

> ... is justified not by *a priori* proofs, but by the fact that they are well-confirmed by experience[15]

---

[14][Gillies, 2000], p. 45.
[15]For Kitcher, although ([Kitcher, 1984], ch. 6, p. 108):

> ... arithmetical truths owe their truth (at one level) to the operations we perform

it is, nevertheless, the case that (ibid.):

> ... arithmetic describes those structural features of the world in virtue of which we are able to segregate and recombine objects ...

But if, as Kitcher says, '... arithmetic describes those structural features of the world in virtue of which we are able to segregate and recombine objects ... ', then arithmetical laws are *a posteriori* true, namely, they are true in virtue of what the world happens to be like.

it follows that Kitcher's and Gillies's philosophies of mathematics fall on the second horn of the dilemma.

Such a situation calls for a re-thinking of the empiricist's position, which will take place through a discussion of some of Lakatos's contributions to epistemology.

6 Quasi-empiricism

Lakatos's contributions to the philosophy of mathematics originate within the context of the traditional epistemological debate which aims at resolving the sceptic's challenge to the idea that there is such a thing as mathematical knowledge.

The traditional epistemological debate started at the time when it became apparent that Kant'sphilosophy of mathematics is untenable.

As is well known, for Kant, there is mathematical knowledge and this is characterized by the fact that, according to Kant, mathematical statements (judgments) are synthetic *a priori*, whereas the statements belonging to physics are synthetic *a posteriori*.

However, the discovery of the non-Euclidean geometries delivered a tremendous blow to Kant's views concerning mathematics in general, and mathematical knowledge in particular, giving life to the sceptic's challenge.

Several, and contrasting, were the solutions offered to the newly open problems of determining whether or not there is mathematical knowledge, and what kind of knowledge this is; and, with regard to questions of methodology, it is particularly important to notice that the general idea that many philosophers had about how to approach the problems above — and, in particular, the latter — remained the Kantian one of determining the nature of mathematical statements (judgments) by means of the distinctions: analytic/synthetic and *a priori*/*a posteriori*.

The most serious attempts made in this direction were, on the one hand, the traditional programmes in the foundations of mathematics: Logicism, Hilbert's programme, Intuitionism; and, on the other, Mill's philosophy of mathematics which considers mathematics as based on experience.

Now, two are the basic ideas from which much of Lakatos's reflection on mathematics proceeds. The first is that, since the traditional programmes in the foundations of mathematics, and Mill's philosophy of mathematics, both failed to establish the certainty of mathematical theories, they also failed to dispel the sceptic's doubts about mathematical knowledge.

The second is that, if we examine the process according to which mathematical knowledge grows, we will realize that this is a process which we do not find in axiomatized mathematical theories, but rather in the twists and turns of informal mathematics; and that such a process evolves according

6. QUASI-EMPIRICISM

to the Popperian model of conjecture and refutation.

A very important outcome of Lakatos's work on the failure of the traditional programmes in the foundations of mathematics, and of his views concerning the untenability of an empiricist philosophy of mathematics, is the distinction he draws between Euclidean and quasi-empirical theories. A Euclidean theory is for him[16]

> ...a deductive system with an indubitable truth-injection at the top (a finite conjunction of axioms) — so that truth, flowing down from the top through the safe truth-preserving channels of valid inferences, inundates the whole system.

In other words, in a Euclidean theory the 'truth-flow' goes from self-evidently true axioms to the mathematical statements which can be deduced from them using sound rules of derivation.

On the other hand, for Lakatos, a quasi-empirical theory is a deductive system in which:[17]

> The important logical flow ...is not the transmission of truth but rather the retransmission of *falsity* — from special theorems at the bottom ('basic statements') up towards the set of axioms.

We must notice here that the distinction Euclidean/quasi-empirical applies, for Lakatos, to *any* deductive system,[18] and is meant to distinguish between different '...pattern(s) of truth value flow in the system'.[19] What these different patterns of truth-value flow show is that in a Euclidean system logic is, as he says, an *organon of proof*, whereas in quasi-empirical systems logic is an *organon of criticism*. Indeed, the retransmission of falsity is a very strong way of criticising/refuting some of the axioms of the deductive system on the basis of the time-honoured logical principle

$$A \to B, \neg B \models \neg A.$$

A consequence of this is that[20]

> A Euclidean theory may be claimed to be true; [whereas] a quasi-empirical theory — at best — to be well-corroborated, but always conjectural. [The italics are mine.]
> Also, in a Euclidean theory the true basic statements at the 'top' of the deductive

---

[16] See [Lakatos, 1967a], p. 28.
[17] See [Lakatos, 1967a], §2, p. 28.
[18] [Lakatos, 1967a], §2, p. 29:

> ...a theory which is quasi-empirical in my sense may be either empirical or non-empirical in the usual sense: it is empirical only if its basic theorems are spatio-temporally singular basic statements whose truth values are decided by the time-honoured but unwritten code of the experimental scientist.

[19] [Lakatos, 1967a], §2, p. 29.
[20] [Lakatos, 1967a], pp. 28-29.

system (usually called 'axioms') *prove*, as it were, the rest of the system; in a quasi-empirical theory the (true) basic statements are *explained* by the rest of the system.

One of the most central features of the Euclidean/quasi-empirical distinction is that, according to Lakatos, through the study of the history of mathematics, it is possible to determine whether a given mathematical theory T is Euclidean or quasi-empirical; and, in case T is quasi-empirical, how we '...turn our guesses [mathematical conjectures] into criticizable [refutable] ones, and criticize [refute] and improve them.'[21]

This, of course, would not directly address the old question of the sceptic about mathematical knowledge, but would provide a satisfactory reply to a new question: How do you improve your guesses? The latter question is, for Lakatos, of paramount importance, because it lies at the root of the logic of mathematical discovery in Popper's sense, i.e., it lies at the root of the theory of mathematical method as what (the method) characterizes mathematics more than the logical structure of mathematical statements and theories.

In concluding this section, we need to notice that a resolution of the debate between Euclideans and quasi-empiricists would have dramatic consequences for the philosophy of mathematics.

In fact, if someone were to show that mathematical theories are Euclidean, a case could still be made in favour of the traditional view of mathematics as the only true science. But if, instead, it could be convincingly argued that mathematical theories are quasi-empirical, we should then give up the traditional view in favour of a more modest picture of mathematics, a picture in which fallibility looms large.

7 Is mathematics quasi-empirical?

For Lakatos mathematics is quasi-empirical, but[22]

> If mathematics and science are both quasi-empirical, the crucial difference between them, if any, must be in the nature of their 'basic statements', or 'potential falsifiers'. The 'nature' of a quasi-empirical theory is decided by the nature of the truth value injections into its potential falsifiers. Now nobody will claim that mathematics is empirical in the sense that its potential falsifiers are singular spatio-temporal statements. But then what is the nature of mathematics? Or, what is the nature of the potential falsifiers of mathematical theories?

According to Lakatos, the quasi-empirical mathematical theories are formal systems which we obtain by formalizing informal theories. And the

---
[21][Lakatos, 1962], §1, p. 10.
[22][Lakatos, 1967a], §4, p. 35.

7. IS MATHEMATICS QUASI-EMPIRICAL? 205

potential falsifiers of these formal systems are eventual contradictions provable within them and, more interestingly, some of the theorems provable within the boiling magma of the corresponding informal theories:[23]

> ...if we insist that a formal theory should be the formalization of some informal theory, then a formal theory may be said to be 'refuted' if one of its theorems is negated by the corresponding theorem of the informal theory. One could call such an informal theorem a *heuristic falsifier* of the formal theory.

I shall call 'strongly fallible' those quasi-empirical mathematical theories which admit of both the above mentioned falsifiers. This is in view of the fact that their refutation is provided 'head on' either by contradictions or by the existence of *heuristic falsifiers*.

It is important to notice that if we accept Lakatos's characterization of a quasi-empirical mathematical theory and of what counts as a falsifier for it, we have that the rules of derivation adopted by such a theory remain unchanged with regard to mathematical practice, and that the refutation of a given formal system is not obtained simply through proof-theoretical considerations, but also by comparing the formal system with the reality represented by the informal theory.

Lakatos's ideas, although very stimulating, present us with a number of problems. In first place, for Lakatos, the distinction between formal and informal mathematics is a distinction in which informal mathematics performs the rôle of some kind of reality, which is the object of description of formal mathematics. That this is not an entirely satisfactory representation of the relationship existing between formal and informal mathematics is shown by the consideration that whereas reality is incorrigible by definition, informal mathematics is not. Informal mathematics is often corrected by the findings and procedures of formal mathematics.

Indeed, for example, informal, or naïve, set theory had to be modified by axiomatic versions of set theory as a consequence of the contradictions, and the introduction of the ϵ, δ-proofs in analysis was the consequence of Cauchy's and Weierstrass's concern and investigations about rigour.[24]

Furthermore, informal mathematics *is* mathematics nevertheless and as such presents us with a number of questions which are common with the formal version of it, e.g., 'In what sense is a mathematical proposition true (or false)?', 'Can we talk about refutation of mathematical theories?', etc.

Finally, the fact that whenever a particular mathematical theory is attacked by contradictions mathematicians do not simply throw it away, but devise appropriate axiom-systems within which those contradictions cannot

---

[23] See [Lakatos, 1967a], p. 36.
[24] See [Grabiner, 1981].

be reproduced, ought to show that it does not make much sense to talk about *logical* or *heuristic falsifiers* for mathematical theories.[25]

Therefore, if mathematical theories are falsifiable at all, this can only mean that a mathematical theory T can be discarded in favour of another theory T'. In the following section I will discuss the case-history of Euclidean geometry in which we have the occurrence of the phenomenon described above.

8 Euclidean geometry: a case-history (I)

Euclid begins Book I of the *Elements* distinguishing between three types of different notions: definitions (*hóroi*),[26] postulates (*aitémata*)[27] and common notions (*koinài énnoiai*).[28] If we analyse Euclid's propositions falling under the three above mentioned categories, we realise that: the group of definitions establishes the meaning of basic geometrical terms; that of postulates characterises the possibility of carrying out certain operations between geometrical entities already defined (postulates 1–3), and posits properties and relations of geometrical entities (postulates 4–5); whereas that of common notions provides *general* criteria of equality (and inequality), i.e., no mention of a geometrical or arithmetical context is made.

We can carry out a further grouping of the notions produced by Euclid into two different classes: i) geometrical (definitions and postulates) and ii) general (common notions).

---

[25] Perhaps the fact that Lakatos speaks here of logical or heuristic falsifiers is simply evidence of a conception of falsifiability which had not yet evolved into that more interesting notion of *sophisticated falsifiability* he elaborated as a consequence of his investigations in the history and philosophy of the empirical sciences. In other words, had Lakatos had enough time to rewrite his philosophy of mathematics, he would perhaps have removed from it these untenable ideas.

[26] These are statements like 'A **point** is that which has no part', 'A **line** is breadthless length', etc. See [Euclid, *The Elements*], Book I, pp. 153–154.

[27] These are:

1) To draw a straight line from any point to any point. 2) To produce a finite straight line continuously in a straight line. 3) To describe a circle with any centre and distance. 4) That all right angles are equal to one another. 5) That, if a straight line falling on to straight lines make the interior angles on the same side less than two right angles, the two straight lines, if produced indefinitely, meet on that side on which are the angles less than the two right angles.

See [Euclid, *The Elements*], Book I, pp. 154–155.

[28] These are:

1) Things which are equal to the same thing are also equal to one another. 2) If equals be added to equals, the wholes are equal. 3) If equals be subtracted from equals, the remainders are equal. 4) Things which coincide with one another are equal to one another. 5) The whole is greater than the part.

See [Euclid, *The Elements*], Book I, p. 155.

8. EUCLIDEAN GEOMETRY: A CASE-HISTORY (I)

According to Heath, what Euclid means by 'common notions' is precisely what Aristotle means by 'axiom'.[29] For Aristotle axioms are universal and self-evidently true principles which are presupposed by all kinds of learning and attempts to provide justifications for holding certain beliefs.[30]

Axioms *must* be self-evidently true for Aristotle, because — quite apart from the consideration already made in the second quotation given in footnote 30 — if this were not the case we would then have to have an infinite regress within the set of axioms which would threaten the very concept of proof.

In fact, if a given axiom A is not self-evidently true then the justification for asserting that A is true must be provided by a demonstration. However, according to Aristotle, in any demonstration we presuppose and/or use axioms. This would in particular be true of a proof of A. But since the axioms we have used in the proof of A are not self-evidently true, we need to provide a proof of them, which in turn will involve other axioms, etc. *ad infinitum*. Therefore, if axioms are not self-evidently true statements, the very concept of proof collapses.

The idea that within the Euclidean tradition axioms were taken to be self-evidently true statements is not shared by all the historians of mathematics.

---

[29]The following text occurs in [Euclid, *The Elements*], Introduction, Ch. IX, §3, p. 124:

> On the whole I think it is from Aristotle that we get the best idea of what Euclid understood by a postulate and an axiom or common notion. Thus Aristotle's account of an axiom as a principle common to all sciences, which is self-evident, though incapable of proof, agrees sufficiently with the contents of Euclid's *common notions* as reduced to five in the most recent text (not omitting the fourth, that "things which coincide are equal to one another").

[30]*Axioms are universal*: see [Aristotle, *Posterior Analytics*], 72^a 15–19:

> An immediate deductive principle I call a posit if one cannot prove it but it is not necessary for anyone who is to learn anything to grasp it [this definition fits very well with the Euclidean notion of postulates]; and one which it is necessary for anyone who is going to learn anything whatever to grasp, [I call] an axiom (for there are some such things); for we are accustomed to use this name especially of such things [this fits extremely well the Euclidean concept of common notion, i.e. notion which is common to the activity of grasping anything whatever].

and again in [Aristotle, *Topics*], $\theta 5$, 159^a 28–30. *Axioms are self-evidently true*: see [Aristotle, *Posterior Analytics*]:

> Demonstration starts from principles which must be admitted. It is no use objecting that a man can admit and refuse to admit whatever he likes. He may of course *say* what he likes, but there are some things which once set before him, he cannot help believing [this passage speaks very strongly in favour of the concept of axiom as a self-evidently true statement]; and demonstration, being concerned with teaching the truth, is directed to what a learner actually believes and not to what he merely says.

Szabó, for instance, argues that axioms were nothing but *requests* made to a partner in conversation to accept a particular assertion as the starting point of a discussion.[31]

He argues this point by considering a particular way in which the word *axíoma* was used — as meaning *request* — and, above all, by drawing a relationship, in the form of a story, between the style of arguing adopted by Eleatic philosophers and that present in the *Elements*.

For Szabó '... Euclid's mathematical "principles" [i.e. definitions, postulates and axioms] were an adaptation of the dialectician's "hypotheses"'[32] and, in his view, an hypothesis was '... a strong initial assertion, which was never proved, but accepted as true without proof'.[33]

First of all it must be said that the representation of a highly specialized subject, such as Euclidean geometry had become at the time when the *Elements* were written, as an extension of a philosophical discussion is very reductive, to say the least.

Euclid himself, and many others after him, were professional mathematicians who dedicated all their time to the development of their subject, a subject characterised by principles, techniques, etc. which had, already for some time, been firmly *founded* not on the basis of a set of conventions such as could be those regulating chess or tennis, but on statements which were commonly *held* to be true and not simply accepted as such for the sake of argument.

Secondly, to establish the meaning of a word such as *axíoma* in relation to its use in mathematics we need to examine texts in which the word occurs in discussions concerning mathematics and/or knowledge. Such relevant occurrences of this word are mainly in Aristotelian texts — Euclid used the expression 'common notions' — and, as we have already seen, Aristotle's conception of justification is *foundationalist* in character. It is a conception based on the idea that any logically compelling justification of a given statement must ultimately appeal to statements which are true and whose justification does not depend on any other statement.

In connection with this discussion, it is extremely important to notice that Aristotle's foundationalist interpretation of Euclidean geometry was not challenged for many centuries. The Euclidean-Aristotelian identification of axioms with common notions in the understanding of such statements as

---

[31]

The Greek word 'axioma' originally meant only '*request*'; one partner *requested* the other to accept his assertion as the starting point of the debate. Euclid's much-discussed axiom, '*The whole is greater than the part*', was also such a request.

[Szabo, 1967], §3, in [Lakatos, 1967], p. 8.

[32] Ibid., p. 6.

[33] Ibid., p. 5.

8. EUCLIDEAN GEOMETRY: A CASE-HISTORY (I)

being self-evidently true statements has been very influential in the history of philosophy and mathematics. Descartes, for one, shared this opinion[34] which was also common to Kant.[35] Proclus himself agrees with this interpretation of the notion of axiom and has a great respect for Aristotle who he calls 'the inspired Aristotle' (*o daimónios Aristotéles*).[36]

---

[34] See [Descartes, 1644], Part One, §§49–50, in [Cottingham et alii, 1987], vol. I, p. 209:

> ... when we recognise that it is impossible for anything to come from nothing, the proposition "Nothing comes from nothing" is regarded not as a really existing thing, or even as a mode of a thing, but as an eternal truth which resides within our mind. Such truths are termed common notions or axioms. The following are examples of this class: *It is impossible for the same thing to be and not to be at the same time*; *What is done cannot be undone*; *He who thinks cannot but exist while he thinks*; and countless others. It would not be easy to draw up a list of all of them; but nonetheless we cannot fail to know them when the occasion for thinking about them arises, provided that we are not blinded by preconceived opinions ... In the case of these common notions, there is no doubt that they are capable of being clearly and distinctly perceived; for otherwise they would not properly be called common notions. But some of them do not really have an equal claim to be called 'common' among all people, since they are not equally well perceived by everyone. This is not, I think, because one man's faculty of knowledge extends more widely than another's, but because the common notions are in conflict with the preconceived opinions of some people who, as a result, cannot easily grasp them. But the selfsame notions are perceived with the utmost clarity by other people who are free from such preconceived opinions.

[35] [Kant, 1787], II Transcendental Doctrine of Method, Ch. I, §1, p. 589:

> 2. *Axioms*. — These, in so far as they are immediately certain, are synthetic *a priori* principles. Now one concept cannot be combined with another synthetically and also at the same time immediately, since, to be able to pass beyond either concept, a third something is required to mediate our knowledge. Accordingly, since philosophy is simply what reason knows by means of concepts, no principle deserving the name of an axiom is to be found in it. Mathematics, on the other hand, can have axioms, since by means of the construction of concepts in the intuition of the object it can combine the predicates of the object both *a priori* and immediately, as, for instance, in the proposition that three points always lie in a plane. But a synthetic principle derived from concepts alone can never be immediately certain, for instance, the proposition that everything which happens has a cause. Here I must look round for a third something, namely, the condition of time-determination in an experience; I cannot obtain knowledge of such a principle directly and immediately from the concepts alone. Discursive principles are therefore quite different from intuitive principles, that is, from axioms; and always require a deduction. Axioms, on the other hand, require no such deduction, and for the same reason are evident — a claim which the philosophical principles can never advance, however great their certainty.

[36] The following quotation shows in a very clear way that Proclus's conception of axiom coincides with that of Aristotle:

> When a proposition that is to be accepted into the rank of first principles is something both known to the learner and credible in itself (*kath'autò pistòn*), such a proposition is an axiom: for example, that things equal to the same thing are equal

9 Euclidean geometry: a case-history (II)

If we now turn to a more modern treatise on geometry, say, Hilbert's *The Foundations of Geometry* (**FG**),[37] we realise that a great deal of changes have taken place with regard to the *Elements* (**E**).

First of all, we find that, in contrast with **E**, where we have definitions, postulates, and axioms, in **FG** primitive notions and axioms are sufficient to characterize the geometrical system.

However, since the primitive notions of a mathematical theory T are not defined in terms of other notions belonging to T, it is legitimate to ask how we can establish the meaning of the primitive notions of **FG**.

For Hilbert, the meaning of the primitive notions of **FG** is not given by intuition, but by the axioms in which the fundamental relations involving these notions are stated. Following the current terminology, I am going to call 'implicit definitions' the sets of axioms of **FG** which establish the meaning of the primitive notions of the system. Explicit definitions are, instead, relegated by Hilbert to the function of abbreviating expressions.

From what has just been said follows that Hilbert does not accept the traditional distinction between axioms and definitions. In fact, according to such a distinction, a distinction which was upheld by Frege in one of his letters to Hilbert on these matters,[38] axioms are genuine statements and, therefore, have a truth-value, but definitions have not. For, since definitions simply establish the meaning of otherwise meaningless expressions, they are not conceivable as genuine statements the truth of which needs either to be proved or to be elucidated somehow.

Furthermore, Frege says,[39] since axioms, principles, and theorems cannot contain terms and symbols whose meaning has not been fixed in advance, because otherwise we might have doubts about which thoughts are expressed by them, it follows that axioms, principles, and theorems cannot be used to fix the meaning of terms which occur within them.

Secondly, a point which deserves much attention at this stage of the discussion is that Hilbert's **FG**, far from representing just a more economical or complete or elegant formulation of Euclidean geometry than **E**, actually caused a dramatic change within this branch of mathematics.

Indeed, **FG** contributed to the general trend then present in algebra, and other fields of mathematics, a trend that saw mathematics becoming a

---

to each other. When the student does not have a self-evident (*autópiston*) notion of the assertion proposed but nevertheless posits it and thus concedes the point to his teacher, such an assertion is a hypothesis

[Proclus, *A Commentary*], 76, in [Morrow, 1970], pp. 62–63.
[37] See [Hilbert, 1947].
[38] Letter from Frege to Hilbert, 27 December 1899.
[39] Letter from Frege to Hilbert, 27 December 1899.

9. EUCLIDEAN GEOMETRY: A CASE-HISTORY (II)   211

science of structures. This extremely important feature of **FG** is expressed by the fact that, since **FG** abstracts from the nature of the entities which satisfy its axioms, it follows that **FG** describes a set of relations on an arbitrary domain, i.e., a structure and not specific objects.[40]

Hilbert was fully aware of this aspect of **FG**, so much so that, in his reply to Frege's criticism of his use of 'implicit definitions', he explicitly says that his formulation of Euclidean geometry is applicable to an infinite number of different domains in so far as the fundamental elements of these domains satisfy the axioms of **FG**.

Thirdly, whereas, according to Euclid, the axioms of **E** are self-evidently true, for Hilbert (and Cantor), instead, it is the consistency of the set of axioms of a mathematical theory T the reason for saying that the axioms of T are true and that the entities (implicitly) defined by them exist.

Fourthly, in **FG** Hilbert includes among his six groups of axioms[41] — in the third group — a statement equivalent to Euclid's Postulate n. 5 and versions of Euclid's axioms n. 1 and 2 appear in a *specialised form*, i.e., in a form which refers to *segments* rather than to *things* in general, in Hilbert's congruence axioms 1 and 2,[42] etc.

Fifthly, the number of axioms used by Hilbert to derive Euclidean geometry is 21, whereas Euclid used only 5 axioms and 5 postulates. The reason for this is the well-known fact that the proofs of several theorems of Euclidean geometry presupposed assumptions which had not been made explicit in the set of postulates and axioms offered by Euclid.

The above mentioned examples of changes which have taken place within Euclidean geometry from the time of Euclid to that of Hilbert are very significant, because they point at a deep discontinuity existing between the *Elements* and Hilbert's system **FG**.

The presence of such a profound discontinuity between **E** and **FG** is reinforced by the observation that in **FG** statements which Euclid would have never called 'axioms' — for instance, Hilbert's axiom of parallels — are called 'axioms' by Hilbert and statements that Euclid would have called 'axioms' are not such for Hilbert, an obvious example of this is Euclid's axiom 5 which was shown to be in general false by Cantor's work in set theory.

Moreover, it seems to me that one of the consequences of Hilbert calling

---

[40]The drive towards abstraction in geometry which is so well exemplified in **FG** was largely anticipated by the contributions of some of the members of the Italian school of geometry. See on this [Avellone *et alii*, 2002].

[41]The six groups of axioms set by Hilbert for Euclidean geometry are axioms of: connection, order, parallels, congruence, continuity, completeness. See [Hilbert, 1947] for more detatils.

[42]See [Hilbert, 1947], pp. 12–13.

'axiom' the Euclidean postulate of parallels, in the awareness that such a postulate is false in models of elliptic and hyperbolic geometries, is that the Euclidean-Aristotelian idea of axiom as a self-evidently true statement has already largely faded away.

Indeed, in the post-Euclidean tradition, beginning with the discovery of non-Euclidean geometries, the concept of axiom of a given theory appears to be performing more and more the rôle of an unproved posit gradually losing its position of self-standing, universal, self-evidently true statement.[43]

This type of situation produces, within geometry itself and the way in which geometers look at their subject, a genuine Kuhnian *Gestalt*-shift. The proliferation of different, mutually exclusive, formal systems which have different, and mutually exclusive, axiomatic bases, causes a kind of Copernican revolution within geometry according to which geometrical intuitions and discoveries are no longer bound to revolve around Euclidean geometry.

The very perception of what geometrically relevant relations are changes further with the advent of linear algebra[44] and topology and in all this one of the most important things that we must notice is that the process of liberation and development of geometry, moving along important patterns which suddenly become perspicuous when geometry is seen through certain systems of representation, is only possible because of the change brought about in the notion of axiom.

However, over and above the examples of discontinuity I have examined, examples of discontinuity present in the Hilbertian representation of Euclidean geometry with respect to the original system of Euclid, one of the most important aspects of the *Gestalt*-shift I have mentioned is given by the difference in the understanding of the concept of justification existing between **E** and **FG**.

As we have seen in the previous section, Euclid's concept of justification rested on a *foundationalist* basis, i.e., the justification provided for the assertion of a mathematical statement had to appeal ultimately to statements which were considered to be self-evidently true.

In Hilbert we find a totally different account of justification, an account which then developed into his programme for the foundations of mathematics. Hilbert is a *coherentist*, in other words, for him the assertion of a statement is justified if the statement is proved by means of sound rules of derivation from statements which constitute a consistent whole. At this

---

[43]The reader interested in the history of some of the changes produced in the understanding of important mathematical notions by the introduction of non-Euclidean geometries — in particular with regard to the concept of mathematical truth — will certainly find rewarding the book by J. L. Richards *Mathematical Visions* — see [Richards, 1988], in particular Ch. 2, pp. 61–114.

[44]Concerning the impact of linear algebra on geometry see [Dieudonné, 1968].

9. EUCLIDEAN GEOMETRY: A CASE-HISTORY (II)

stage, it does not matter whether the axioms of a given theory are self-evidently true or not, the only thing that is really important is that the set of axioms of the theory is consistent.

The change operated by Hilbert in the concept of justification had momentous consequences. Its importance did not simply lie in legitimising a multiplicity of axiomatic systems by introducing a principle of tolerance. What Hilbert caused was an important shift in the criteria of acceptability of a mathematical theory.

For Hilbert, once the consistency conditions are satisfied, what becomes of paramount importance in judging whether a mathematical theory is good or not is its fruitfulness and not the appeal to a misguided concept of self-evidence of its axioms.[45] It is extremely important to notice that in saying this I am neither denying that prior to Hilbert *consistency* and *fruitfulness* were among the criteria of choice available to mathematicians nor that, after Hilbert, *simplicity*, *self-evidence*, etc. are still important criteria of choice. What I am asserting is that, as a consequence of Hilbert's coherentist views, the system of *relative* values that the several criteria of choice for a mathematical theory have has changed dramatically.

As the considerations contained in this section clearly show, we have here an example of a situation in which mathematical knowledge does not grow by means of a simple process of accumulation. When we study the fact of the matter carefully enough, there is no doubt that the advent of the Hilbertian version of Euclidean geometry constitutes a growth of knowledge with respect to the system **E**. At the same time, there also is no doubt that such a growth is not cumulative: this has taken place at the expense of the traditional **E**-concepts expressed by the words 'axiom', 'postulate', 'defini-

---

[45][Hilbert, 1926], p. 184:

A careful reader will find that the literature of mathematics is glutted with inanities and absurdities which have had their source in the infinite. For example, we find writers insisting, as though it were a restrictive condition, that in rigorous mathematics only a *finite* number of deductions are admissible in a proof — as if someone had succeeded in making an infinite number of them. Also old objections which we supposed long abandoned still reappear in different forms. For example, the following recently appeared: Although it may be possible to introduce a concept without risk, i.e., without getting contradictions, and even though one can prove that its introduction causes no contradiction to arise, still the introduction of the concept is not thereby justified. Is not this exactly the same objection which was once brought against complex-imaginary numbers when it was said: "True, their use doesn't lead to contradictions. Nevertheless their introduction is unwarranted, for imaginary magnitudes do not exist"? If, apart from proving consistency, the question of the justification of a measure is to have any meaning, it can consist only in ascertaining whether the measure is accompanied by commensurate success. Such success is in fact essential, for in mathematics as elsewhere success is the supreme court to whose decisions everyone submits.

tion', 'justification', 'object of investigation', etc. which have undergone a radical shift in meaning.

Moreover, the very criteria available to the mathematical community to decide whether or not a given mathematical theory is good (and therefore acceptable) have changed in their relative values giving rise in this way to judgements which are in sharp contrast and indeed are incommensurable with those given with the same set of criteria endowed with a different system of weighting.

But, perhaps, one of the most important things which have emerged in the discussions of this section is that, within what we call 'Euclidean geometry', we can distinguish at least two different theories: Euclid's system **E** and Hilbert's **FG**. This, far from being an exception in mathematics, is the norm, and in the next chapter we shall see that these sequences of theories to which we usually refer with terms like 'Euclidean geometry', 'set theory', etc. are actually describable as Lakatosian Mathematical Research Programmes (MRPs). But, of course, before we do so, I must clarify the notion of Lakatosian MRP.

10 Scientific Research Programmes

The concept of Scientific Research Programme (SRP) was originally formulated within the context provided by Lakatos's work in the philosophy of science, and its applicability to the philosophy of mathematics has been supported by some authors[46] and challenged by others.[47] But before I express my opinion on this matter, I need to clarify what Lakatos means by 'Scientific Research Programme'.

A Scientific Research Programme is, in Lakatos's view, a sequence of theories developing from an original hard core according to certain methodological rules. Some of these rules tell us which research directions to avoid: *negative heuristic*; and the others tell us which research directions to take: *positive heuristic*.[48] For instance, in the case of the Newtonian mechanics SRP the hard core is represented, according to Lakatos, by the three laws of motion and by the law of gravitation.

Now, whereas the negative heuristic specifies the hard core of the research programme[49] forbidding researchers to refute assumptions belonging to it,

---

[46]See [Hallett, 1979a] and [Hallett, 1979b].

[47]See [Koetsier, 1991], Ch. IV.

[48][Lakatos, 1983a], §3, p. 47:

The programme consists of methodological rules: some tell us what paths of research to avoid (*negative heuristic*), and others what paths to pursue (*positive heuristic*).

[49][Lakatos, 1983a], §3(b), pp. 49–50:

Research programmes, besides their negative heuristic, are also characterized by

10. SCIENTIFIC RESEARCH PROGRAMMES

the positive heuristic, instead, leads researchers towards the construction of *auxiliary hypotheses* which are useful to eliminate the anomalies resulting from an application of the research programme made in terms of predictions, etc. Lakatos calls such a set of auxiliary hypotheses the *protective belt* of the hard core.[50]

The special status held by the hard core in a given scientific research programme — as a set of prescriptions deemed to be '... "irrefutable" by the methodological decision of its proponents' — is evidence of the non-empirical function of its elements. Such a special status held by the elements of the hard core of an SRP is also evidence of the fact that what is significant for the scientific research programme is something that must be determined only once the SRP has been set up.

If this is correct, the acceptance of a scientific research programme generates a *Gestalt*-shift which affects the selection of features of the perceptual input that are relevant to the explanatory/predictive activity of the scientific research programme.

Moreover, it is important for us to realize that SRPs do not come with sharply delineated distinctions between elements of the hard core and auxiliary hypotheses. Such distinctions between elements of SRPs as those introduced by Lakatos can be recognized only by means of rational reconstructions.

However, at this stage of the discussion I need to emphasize that in the Lakatosian view I am advocating the hard core/protective belt distinction is neither equivalent to nor does it mimic the analytic/synthetic distinction. The reason for this is that what distinguishes the elements of the hard core of an SRP from the elements of the protective belt of the SRP is related neither to whether the elements of the hard core are ampliative judgements or not, nor to whether or not they can be shown to be provable from logic and definitions of logical nature alone; but to the fact that expressions belonging to the hard core of an SRP *have been given* metaphysical status (they are

---

their positive heuristic ...The negative heuristic specifies the 'hard core' of the programme which is 'irrefutable' by the methodological decision of its proponents; the positive heuristic consists of a partially articulated set of suggestions or hints on how to change, develop the 'refutable variants' of the research-programme, how to modify, sophisticate, the 'refutable' protective belt.

[50][Lakatos, 1983a], §3(a), p. 48:

All scientific research programmes may be characterized by their '*hard core*'. The negative heuristic of the programme forbids us to direct the *modus tollens* at this 'hard core'. Instead, we must use our ingenuity to articulate or even invent 'auxiliary hypotheses', which form a *protective belt* around this core, and we must redirect the *modus tollens* to *these*.

irrefutable) by the proponents of the research programme, whereas those belonging to the protective belt of the research programme are considered as ordinary refutable statements.

According to Lakatos, SRPs can be either *progressive* or *degenerating*:[51]

> Let us take a series of theories, $T_1, T_2, T_3, \ldots$ where each subsequent theory results from adding auxiliary clauses to (or from semantical reinterpretations of) the previous theory in order to accommodate some anomaly, each theory having at least as much content as the unrefuted content of its predecessor. Let us say that such a series of theories is *theoretically progressive* (or '*constitutes a theoretically progressive problemshift* [SRP]') if each new theory has some excess empirical content over its predecessor, that is, if it predicts some novel, hitherto unexpected fact. Let us say that a theoretically progressive series of theories is also *empirically progressive* (or '*constitutes an empirically progressive problemshift* [SRP]') if some of this excess empirical content is also corroborated, that is, if each new theory leads us to the actual discovery of some *new fact*. Finally, let us call a problemshift [SRP] *progressive* if it is both theoretically and empirically progressive, and *degenerating* if it is not.

The distinction between *progressive* and *degenerating* SRPs is very important, because it provides a rational criterion of choice between opposing and incompatible SRPs.

Lastly, for Lakatos, the sequence of theories $T, T', \ldots, T''''',$ which characterizes a given progressive SRP, is made of falsifiable theories all of which but the last have been falsified. However, before going any further, I must point out that Lakatos's concept of falsifiability of a theory is very different from Popper's. In fact, for Popper scientific theories are falsifiable in the sense that some of their predictions can be shown to be false by *experimenta crucis* (naïve falsifiability). For Lakatos, instead, a scientific theory is falsifiable only in the sense that it can be superseded by a better theory (sophisticated falsifiability). To avoid any equivocity in the use of the term 'falsifiable', in what follows I shall write *falsifiable* when this term is used in Lakatos's sense. But when is, for Lakatos, a scientific theory superseded by a better one?

According to Lakatos:[52]

> ...a scientific theory T is *falsified* if and only if another theory T' has been proposed with the following characteristics: (1) T' has excess empirical content over T: that is, it predicts *novel* facts, that is, facts improbable in the light of, or even forbidden, by T; (2) T' explains the previous success of T, that is, all the unrefuted content of T is included (within the limits of observational error) in the content of T'; ...(3) some of the excess content of T' is corroborated;

and (4) T' must be able to provide predictions/explanations of events for the predictions/explanations of which T' has *not* been devised (Zahar's criterion).[53]

---
[51][Lakatos, 1983a], §2(c), pp. 33-34.
[52][Lakatos, 1983a], §2(c), p. 32.
[53]See [Lakatos & Zahar, 1983].

10. SCIENTIFIC RESEARCH PROGRAMMES 217

We must notice that if it is true that, for Lakatos, there is a very strong conventional element in the way that an SRP is set up, it is also the case that the *falsifiability* of theories belonging to a given progressive SRP, shows that the criteria adopted by the scientific community to choose or reject a theory belonging to an SRP are preeminently logical rather than psychological or sociological. What I mean by this is that, according to Lakatos, the criteria of choice adopted by the scientific community *ultimately* rest on a rational evaluation of properties of the theories rather than on a mood or a fashion.[54] The same considerations apply to the choice made by the scientific community between two opposing and mutually incompatible SRPs, a choice based on the *progressive/degenerative* distinction.

However, when we turn to Lakatos's definition of *falsifiability*, we must realize that, as I argued in 'Criticism and Growth of mathematical knowledge', in the case of mathematical theories 'excess empirical content' of a theory T' with respect to a theory T must be interpreted as 'results which are provable in T' but not in T'. Condition (2) is obvious, condition (3) is not applicable except in the sense that the theory T', in which it is possible to prove results which are not provable in T, has a model, and condition (4) simply means what it asserts.

The above reinterpretation of the *falsifiability* conditions, which renders them applicable to mathematical theories, shows, on the one hand, the existence of differences between mathematical theories and theories belonging to the empirical sciences, and justifies us in talking about Mathematical Research Programmes (MRPs), rather than about Scientific Research Programmes (SRPs), when dealing with mathematical theories; and suggests, on the other hand, an obvious extension of the *progressive/degenerative* distinction to MRPs, a distinction drawn along the following lines: an MRP is *theoretically progressive* if there are results provable in each new theory belonging to it, which are not provable within its predecessor; an MRP is *factually progressive* if each new theory has a model. An MRP is *progressive* if it is both *theoretically* and *factually progressive* and *degenerating* if it is not.

Now that we know what a Lakatosian SRP is and how we should modify the concept of *falsifiability* to make this applicable to mathematical theories the question to ask is whether mathematical theories, and in particular set

---

[54]Lakatos would then resist positions such as Ernest's *social Constructivism* (see on this [Ernest, 1997]) and Glas's

... 'post-positivist' [view of mathematics which emphasizes] cognitive practices drawing on socially shared and socially transmitted goals, methods, standards and criteria.

in [Glas, 1995], p. 225.

11 Mathematical Research Programmes

theory, are conceivable as Lakatosian Mathematical Research Programmes.

The applicability of the concept of Lakatosian SRP to mathematics, never attempted nor upheld by Lakatos himself, has been questioned by Koetsier in [Koetsier, 1991]. Koetsier's scepticism about the relevance of MSRPs (Methodology of Scientific Research Programmes) to mathematics depends on three main reasons.

The first is that:[55]

> [In mathematics] there is no major difference between, say, a general conjecture and a very special case of it, comparable to the difference between Newton's simplest models [of the solar system] and the empirical facts.

To this it is possible to reply providing two examples in which accepted models of mathematical objects and operations are eventually discarded as inadequate and new ones are adopted in their place.

The first such example is represented by the case of the Newtonian-Leibnizian infinitesimals I discussed in §1.6. I there noticed that the Newtonian-Leibnizian infinitesimals provided a model for the operations of differentiation, integration and for the concept of continuity, and that this model suffered from a number of anomalies which were eventually removed not simply by restricting the use of the concept of infinitesimal in analysis, but through the actual elimination of infinitesimals and the introduction of a new model formulated in terms of limits.

The second example is given by the various models of number produced within Western mathematics. Starting from the 'dots and pebbles' of the Pythagoreans, we then encounter the Euclidean concept of number as the length of a line segment. Continuing our survey of the models of the concept of number offered from antiquity to the present day, we end up eventually our intellectual journey with the latest views on numbers as sets or as positions in structures.

If we place all these models of number in a temporal sequence starting from, say, the Pythagorean one, we realize that such models, very much like the various models of the solar system produced by Newton, keep improving as we move down the line.

The second reason given by Koetsier for the impossibility of applying SRPs to mathematics is that[56]

> In Lakatos's rational reconstructions on the basis of his MSRP there are always rival research programmes concerning a particular aspect of empirical reality. However,

---

[55][Koetsier, 1991], Ch. II, §II.4, p. 69.
[56]Ibid.

11. MATHEMATICAL RESEARCH PROGRAMMES

there seems to be considerably less competition in mathematics than in science, at least at first sight.

The objection above, like the one I examined before it, seems to be rather unfair. In fact, if intuitionistic analysis, RUSS analysis, and classical analysis are conceivable as MRPs then they offer a clear example of rival MRPs. And the fact that we can reproduce such an opposition between an intuitionistic, a RUSS, and a classical MRP for every mathematical theory shows that, after all, also in mathematics there is much competition between rival MRPs.

But, perhaps, someone may object to this example on the grounds that the differences between intuitionistic, RUSS, and classical analysis are so profound that intuitionistic, RUSS, and classical analysis cannot even be conceived as rival theories. To such a person I can reply mentioning one of the many cases of rival MRPs which, within classical mathematics, jostle for supremacy: the opposition beween the structuralist MRP in number theory initiated by Dedekind and more traditional approaches to the subject such as Kronecker's.

The difference between Dedekind's and Kronecker's views on number theory do not simply affect the concept of natural number,[57] but generate several other concepts — that of *ideal*, for instance — which, introduced into number theory to solve particular problems (factorization of numbers), contribute greatly not only to the development of number theory, but also to the rise of abstract algebra.

The third, and last, objection made by Koetsier against the applicability of Lakatosian MRPs to mathematics goes as follows:[58]

> [In contrast with what happens in the empirical sciences] *mathematical theories are* [only] *weakly fallible in the sense that one can never exclude the occurrence of unintended possible interpretations of fundamental notions that require a restriction of universality claims by means of conceptual refinement. Because only the range of validity of theorems is restricted weak fallibility implies far going continuity.*

---

[57] Dedekind has a clearly structuralist view of natural numbers, as shown by the quotation below, whereas Kronecker upholds a view of natural numbers as objects. [Dedekind, 1963b], §73, p. 68:

Definition 11.1. If in the consideration of a simply infinite system N set in order by transformation ϕ we entirely neglect the special character of the elements; simply retaining their distinguishability and taking into account only the relations to one another in which they are placed by the order-setting transformation ϕ, then are these elements called *natural numbers* or *ordinal numbers* or simply *numbers*, and the base-element 1 is called the *base-number* of the *number-series* N. With reference to this freeing the elements from every other content (abstraction) we are justified in calling numbers a free creation of the human mind.

[58] [Koetsier, 1991], Ch. X, §X.2.1, p. 278.

Koetsier's weak fallibility thesis concerning mathematical theories implies that mathematical knowledge grows in a cumulative way. For since the worst that can happen to a mathematical theorem is having 'its validity restricted', as a consequence of what Koetsier calls 'conceptual refinement', neither does the meaning of a theorem change nor is a theorem ever refuted; and, consequently, mathematical knowledge, in contrast with that produced by the empirical sciences, grows in a cumulative way.

Since arguing against Koetsier's weak fallibility thesis requires a careful discussion of case-histories and this chapter is already long enough, I shall attempt to show that Koetsier's weak fallibility thesis is incorrect in the next chapter.

12 Quasi-empiricism, foundations, foundationalism

Having made big promises about giving (in the next chapter) an account of mathematical theories in terms of Lakatosian Mathematical Research Programmes, and having provided ample evidence in favour of the quasi-empirical nature of mathematics, it is important, before bringing this chapter to a close, to ask which, if any, are the consequences of such a view on the study of the foundations of mathematics. In order to tackle this problem I need to clear the decks from sources of possible misunderstanding by providing an elucidation of the terms in play.

If by 'foundations' in the philosophy of mathematics we mean the study of the logic and epistemology of mathematics whose goal '... is to establish once and for all the certitude of mathematical methods,'[59] there is no doubt that, first, what we know as Logicism, Intuitionism, and Hilbert's programme belong to the foundations of mathematics; and, secondly, that Lakatosian quasi-empiricism is opposed to such foundations.

With regard to the first point above, if the logicist is right and there is no boundary line between logic and mathematics (arithmetic, for Frege) the certitude of mathematical (arithmetical) methods would be established by the fact that mathematical (arithmetical) theorems are obtained by means of sound rules of derivation (ultimately) from self-evidently true logical statements.

Moreover, if the intuitionist is right then the certitude of mathematical methods would rely on constructive procedures which originate (ultimately) from 'the fundamental phenomenon of the human intellect', i.e., from [60]

> ... the falling apart of moments of life into qualitatively different parts, to be reunited only while remaining separated by time ... passing by abstracting from its emotional content into the fundamental phenomenon of mathematical thinking, the intuition of the bare two-oneness.

---

[59] [Hilbert, 1926], p. 184.
[60] [Brouwer, 1912], p. 80.

12. QUASI-EMPIRICISM, FOUNDATIONS, FOUNDATIONALISM

Furthermore, if the supporter of Hilbert's programme is right, the certitude of mathematical methods would be established by the fact that mathematical theorems are obtained from a consistent set of axioms by means of sound rules of derivation.

Concerning the second assertion made above about Lakatosian quasi-empiricism being incompatible with foundations, we must consider that the attempt to establish once and for all the certitude of mathematical methods is clearly opposed to the fallible nature of mathematical theories, which is an idea at the very heart of quasi-empiricism in mathematics:[61]

> Why foundations, if they are admittedly subjective? Why not honestly admit mathematical fallibility, and try to defend the dignity of *fallible* knowledge from cynical scepticism, rather than delude ourselves that we shall be able to mend invisibly the latest tear in the fabric of our 'ultimate' intuitions?

One of the most important consequences of Logicism, Intuitionism and Hilbert's programme being, on the one hand, *foundational* and, on the other hand, incompatible with quasi-empiricism is that the failure of these programmes to achieve their goals provides indirect confirmation of the correctness of quasi-empiricism.

However, if, in contrast with the definition given above, by 'foundations' in the philosophy of mathematics we mean the study of the logic and epistemology of mathematics whose aim is the unification of mathematical theories, for example, by means of arguments to the effect that mathematics is, at the root, a science of sets or of structures, etc. and if the theories in terms of which the unification is carried out are MRPs, there is no reason why foundations ought to be incompatible with quasi-empiricism. Indeed, as we shall see in Chapter 7, from the fact that set theory is an MRP follows that set theory is quasi-empirical.

Moreover, should every such attempt at the unification of mathematical theories turn out to be just another myth, because of the existence of at least two different theories in terms of which to carry out successfully such a unification, e.g., set theory and category theory, this would not show that there is something wrong with saying that mathematical theories are quasi-empirical, quite the contrary. If such a situation were to obtain, mathematical theories would have a further property in common with empirical theories[62]

> For the chief characteristic of empirical science is that for each theory there are usually alternatives in the field, or at least alternatives struggling to be born.

Therefore, given the two meanings of the term 'foundations' in the order in which they have been discussed in this section, we can conclude that

---

[61][Lakatos, 1962], §3, p. 23.
[62][Putnam, 1967], p. 302.

quasi-empiricism in the philosophy of mathematics, although incompatible with the former meaning, turns out to be independent of the latter. (The meaning attributed by Lakatos to the term 'foundations' is the former.)

Another term which becomes often entangled in discussions connected with quasi-empiricism is 'foundationalism'. If by 'foundationalism' we mean the view of justification according to which any logically compelling justification for the assertion of a given statement must ultimately appeal to statements which are true and whose justification does not depend on any other statement, it follows that foundationalism is incompatible with a quasi-empiricist view of justification for the assertion of a mathematical statement.

In fact, if, as the quasi-empiricist believes, mathematical theories are conjectural, the justification for the assertion of a mathematical statement belonging to such theories cannot in general be something which 'ultimately appeals to statements which are true and whose justification does not depend on any other statement'. For, if this were the case then each correctly asserted mathematical statement would be *proved* and, consequently, mathematical theories, in contrast with the quasi-empirical view, would not be fallible.

CHAPTER 7

A RATIONAL RECONSTRUCTION OF CANTOR-ZERMELO SET THEORY

1 Introduction

At the heart of Chapter 6 there is the thesis that mathematics is a quasi-empirical science. And the correctness of this thesis has been confirmed by the case-history centred on Euclidean geometry.

However, since there is more to mathematics than Euclidean geometry, to show that the central thesis of Chapter 6 is correct either I need to take the long and winding road of providing, for every mathematical theory T, a rational reconstruction of the historical development of T from which it should emerge that T is quasi-empirical or I have to find a general argument to the same effect.

Although the first route to justifying the view that mathematics is a quasi-empirical science promises to be very rewarding in terms of the richness of the information to be produced concerning the peculiarities of the historical development of every single mathematical theory, the type of work involved in following it is such that it can neither find sufficient room in a single book nor can it be carried out by a single man.

What is, therefore, left for me to do in this last chapter of the book is to proceed along the second possible route indicated above: to give a general argument in support of the idea that mathematics is a quasi-empirical science.

The argument in support of the idea that mathematics is a quasi-empirical science that I am going to offer in this chapter will first establish that Cantor-Zermelo set theory is a Lakatosian Mathematical Research Programme. This is a very important stepping stone on the way to achieving my goal, because, as we know from Chapter 6 where I discussed the notion of Lakatosian MRP, if Cantor-Zermelo set theory is a Lakatosian MRP, this implies that Cantor-Zermelo set theory is quasi-empirical.

The second, and decisive, part of the argument consists in the realization that, since it is possible to reduce all mathematical theories to Cantor-Zermelo set theory, if Cantor-Zermelo set theory is quasi-empirical we are then entitled to believe that mathematics is a quasi-empirical science.

While developing this line of argument, I shall also produce evidence in favour of the idea that whereas Cantor-Zermelo set theory is a MRP, its main anti-realist rival research programme — constructivism — presently finds itself in a pre-MRP stage of development. I shall then exploit these features of Cantor-Zermelo set theory and constructivism in an argument to the effect that set-theoretical realism is, so far, the metaphysical view of mathematics which receives the strongest support/confirmation from the history of mathematics.

At this point someone might ask what is so special about Cantor-Zermelo set theory. Indeed, after Cantor's contributions to set theory, which crystallized in the informal system known as naïve set theory (**NST**), there has been a proliferation of axiomatic systems, Zermelo's **Z** and **ZFC**, von Neumann's and Bernay's **VNB**, Morse's and Kelley's **MK** being only some of the best known of these.

There are several reasons at the root of my decision to concentrate on Cantor-Zermelo set theory; I am here going to mention only three of them.

The first is that, since Zermelo's ideas concerning the axiomatization of set theory gave rise, through the contributions of several mathematicians, among whom is prominent Fraenkel, to an explicit plurality of axiomatic systems which supersede each other (**Z** is superseded by **ZFC**), it seems to me that Zermelo-Fraenkel set theory presents us with a *prima facie* case for the application of the concept of Lakatosian MRP.

The second is that **ZFC** is the current main-stream axiomatic system for set theory used by most set theorists and other mathematicians alike.

The third reason refers to a cluster of technical considerations relating to important features of Cantor's **NST**, which have been carefully preserved in **Z** and **ZFC**, but not in other formal systems.

The chapter shall begin with a discussion of the origins of set theory from which will emerge elements of the hard core of the Cantor-Zermelo MRP. I shall then concentrate on the study of **NST**, which will provide a sharper chracterization of the hard core of the Cantor-Zermelo MRP and useful information concerning some of the elements of the protective belt of **NST**. At this point Zermelo's systems **Z** and **ZFC** will be examined bringing out the *progressive* nature of the Cantor-Zermelo MRP.

The chapter will then proceed with an analysis of constructivism which will reveal that this anti-realist research programme is still in a pre-MRP stage of development, and will end with a new argument in favour of mathematical realism.

2 The pre-history: Cauchy analysis

Set theory was developed by G. Cantor towards the end of the nineteenth century, but some of its deepest themes were at the very centre of important changes in analysis which led from the calculus of infinitesimals of Newton and Leibniz to the post-Weierstrassian system.

Before discussing whether or not set theory is conceivable as a Lakatosian MRP, it is important to turn briefly to the consideration of some such changes to cast light on which ideas became central to set theory and on why this happened to be the case. This preliminary discussion is, of course, essential to determine in a non-arbitrary way which concepts belong to the hard core and which to the protective belt of the would-be set theory MRP.

In §1.6 I described one important aspect of the change of theory between Newtonian-Leibnizian analysis (T) and what I might call 'post-Weierstrassian analysis' (T'): the dramatic rejection of infinitesimals and the adoption in their place of the theory of limits.

However, to be faithful to the actual historical development, it must be said that neither were infinitesimals suddenly brushed away nor did the theory of limits appear in its T' form. But post-Weierstrassian analysis was somewhat preceded by what Grattan-Guinness,[1] calls 'Cauchy analysis' (or T^*).[2]

Some of the main characteristics of T^* are: (i) the introduction of the concept of limit which is applied to the definitions of the derivative and of the integral, (ii) the belief in infinitesimals, and (iii) the use of the concept of potential infinity.

There is no doubt that characteristic (i) clearly marks a point of departure of T^* from T, even though '...the system [of differential calculus Cauchy] produced was not at all like the modern treatment'.[3] In particular, if, as Grattan-Guinness says, we consider the definition Cauchy gave of infinitesimal:[4]

> We say that a variable quantity becomes *infinitely small*, when its numerical value decreases indefinitely so as to converge towards the limit 0

---

[1] See [Grattan-Guinness, 1980], Ch. 3, §§3.6–3.8.

[2] As a matter of fact 'Cauchy analysis' shared many characteristics and aims in common with Bolzano's programme to make analysis more rigorous than it had been until then, even though Bolzano's ideas, in contrast with Cauchy's, received very little attention.

Indeed the similarities existing between the views of Cauchy and Bolzano are so striking that they have prompted some scholars to assert that they are not explainable simply in terms of simultaneous discoveries. See on this [Grattan-Guinness, 1979], [Freudenthal, 1971], and [Grabiner, 1984].

[3] [Grattan-Guinness, 1980], p. 111.

[4] The quotation from Cauchy is to be found in [Grattan-Guinness, 1980], Ch. 3, §3.6, p. 110.

we realize that Cauchy's concept of variable greatly differs from that adopted in post-Weierstrassian analysis. For Cauchy the word 'variable' refers to a variable quantity, whereas in the post-Weierstrassian tradition the word 'variable' is used to designate any element of a domain.

Therefore, whereas Cauchy's notion of variable is compatible with a view of infinity as potential, because the variable quantity varies according to a process, the post-Weierstrassian view of a variable is linked with the fact that the domain is actually given.

These considerations are very important for us, because in post-Weierstrassian analysis (1) the concept of 'domain' is nothing but a proto-concept of set; and because (2) since in post-Weierstrassian analysis an infinite domain is treated as actually given, we have here the coming about of a mathematically legitimate use of the concept of actual infinity.

Another, but connected, influence on the concepts of set and actual infinity coming to the forefront of research in mathematical analysis was excercized by the development of the theory of functions. I shall briefly discuss this topic in the following section.

3 The pre-history: the theory of functions

As is well known, Dirichlet was led to the definition of the concept of single-valued arbitrary function as a consequence of his investigations of the problem of the representability of functions by means of trigonometrical series.

If by 'trigonometrical series' we mean a series of the form

$$\frac{a_0}{2} + \sum_{n=1}^{\infty}(a_n \cos nx + b_n \sin nx)$$

according to Hobson[5]

> Fourier shewed, in a variety of special cases, that a function $f(x)$ is representable for values of x between $-\pi$ and π, by the series [above] where
>
> $$a_n = \frac{1}{\pi}\int_{-\pi}^{\pi} f(x)\cos nx\, dx, \quad b_n = \frac{1}{\pi}\int_{-\pi}^{\pi} f(x)\sin nx\, dx,$$
>
> ...Although Fourier attained to correct views as to the nature of the convergence of the infinite series he employed, he did not give any complete general proof that the series in the general case actually converges to the value of the function; he indicates however a process of verification of such convergence which was not actually carried out until Dirichlet took up the subject.

Dirichlet's definition of single-valued arbitrary function and his investigations concerning the representability problem, together with the contributions given by Riemann, Hankel, and others, lay the foundations of the

---

[5][Hobson, 1921], Vol. II, Ch. VIII, §316, pp. 480–481.

4. THE HARD CORE OF A NEW MRP (1)

theory of functions of a real variable, which has since then become one of the most important branches of analysis. But what is the relevance of the theory of functions of a real variable to the concepts of actual infinity and of set?

First of all, the almost universal acceptance of Dirichlet's extensional conception of a function as being determined by its graph, in place of the traditional view of a function given in terms of an operation, made unavoidable, and extremely pervasive, the concept of actual infinity within analysis.

Secondly, as Hobson explains in what follows, set theory and point-set topology represent the ideal setting for the theory of functions of a real variable, because since[6]

> The object to be attained by the theory of functions of a real variable consists ...largely in the precise formulation of necessary and sufficient conditions for the validity of the limiting processes of Analysis. A necessary requisite in such formulation is a language descriptive of particular aggregates of values of the variable, in relation to which functions possess definite peculiarities. This language is provided by the Theory of Sets of Points, also known, in its general aspects, as the theory of Aggregates, which contains an analysis of the peculiarities of structure and of distribution in the field of the variable which such sets may possess.

Now that I have examined some of the factors that, as a consequence of the internal development of analysis, caused the concepts of actual infinity and of set to come to the forefront of research in this branch of mathematics, it is time to see how set theory began.

4 The hard core of a new MRP (1)

As the history of mathematics reveals, set theory originated from two main sources. The first was Cantor's investigations within analysis concerning whether[7]

> ...given an arbitrary function represented by a trigonometric series, ...the representation [is] necessarily unique

And the second was his proof that $\mathbb{Q}$ is denumerable and $\mathbb{R}$ is not denumerable.

In fact, Cantor's research on the representation problem led him to introduce the crucial notion of *derived set*. This notion is very important for the birth of set theory for two reasons. The first is that here the notion of set explicitly plays a central rôle. And the second is as follows. Given a set

---

[6][Hobson, 1921], vol. I, Preface to the First Edition, p. vi.
[7][Dauben, 1990], Ch. 2, p. 30. Giving an affirmative answer to the problem formulated in the quotation above is equivalent to proving that: if there are two trigonometrical series both of which represent the same function, the trigonometrical series obtained by subtracting the first series from the second must have all its coefficients equal to 0.

of points on the real line P, we call the set $P^{(1)}$ 'the first derived point-set of P', where
$$P^{(1)} = \{x : x \text{ is a limit point of } P\}.^8$$
For example, if
$$P = \{x \,:\, x = 1 + \frac{1}{n}, \text{ for } n \in \mathbb{N} \text{ and } n \neq 0\},$$
then $P^{(1)} = \{1\}$.

If $P^{(1)}$ is a finite set, as in the example above, then $P^{(1)}$ will *not* have a limit point and, therefore, $P^{(2)} = \emptyset$. But if $P^{(1)}$ is an infinite set and there exists a closed interval I_1 of reals such that $P^{(1)} \subseteq I_1$, by the Bolzano-Weierstrass theorem, $P^{(1)}$ will have a limit point and, therefore, $P^{(2)}$ will not be empty.

Cantor argues in [Cantor, 1879] that the ideas expressed above enable us to distinguish between two types of sets of reals: (1) *sets of the first type*, i.e., those such that there exists an $n \in \mathbb{N}$ such that $P^{(n)} = \emptyset$; and (2) *sets of the second type*, i.e., those for which $P^{(\infty)} \neq \emptyset$. An example of set of the second type is the set P of all the irrational numbers belonging to the interval $[0, 1]$. In this case $P^{(1)}$ will be the whole interval and
$$P^{(1)} = P^{(2)} = \cdots = P^{(\infty)} \neq \emptyset.$$

An even simpler example is given by any set of reals P such that $P = P^{(1)}$.

Within the collection of sets of the first type, we can draw a further distinction introducing the concept of 'set of the n-th species', for $1 \leq n$. We say that a set of reals of the first type P is of the n-th species iff $P^{(n)} \neq \emptyset$ and $P^{(n+1)} = \emptyset$.

We must now consider, as Cantor did in [Cantor, 1879], that, given an interval of finite length P of the second type, if $P^{(\infty)}$ contains an infinite number of points, since
$$P^{(\infty)} \subseteq \cdots \subseteq P^{(2)} \subseteq P^{(1)},$$
there exists a closed interval I_∞ such that $P^{(\infty)} \subseteq I_\infty$ (we can trivially put $I_\infty = I_1$). Therefore, $P^{(\infty)}$ must have a limit point (Bolzano-Weierstrass theorem again), and the operation of derivation can be applied to $P^{(\infty)}$ obtaining
$$P^{(\infty+1)} = (P^{(\infty)})^{(1)},$$

---

[8]

Definition 4.1. If A is a subset of $\mathbb{R}$, a $p \in \mathbb{R}$ is a limit point of A iff for any open set $G \subseteq \mathbb{R}$, if $p \in G$ then $A \cap (G \backslash \{p\}) \neq \emptyset$.

4. THE HARD CORE OF A NEW MRP (1)

and so on.

A very important thing that we must notice at this point is that the exponents

$$\infty, \infty+1, \infty+2, \ldots$$

which appear in the sequence of derived sets of P:

$$P^{(1)}, P^{(2)}, \ldots, P^{(\infty)}, P^{(\infty+1)}, \ldots$$

represent an extension of the number field beyond the finite numbers

$$1, 2, \ldots$$

an extension obtained through a transfinite iteration of the operation of derivation.

The second source from which set theory originated was Cantor's theory of the continuum, which showed $\mathbb{R}$ to be much richer than $\mathbb{Q}$ (no *gaps* in $\mathbb{R}$), making him suspect that the number of elements of $\mathbb{R}$ might be greater than the number of elements of $\mathbb{Q}$.

The above mentioned result concerning the difference in numerosity existing between $\mathbb{Q}$ and $\mathbb{R}$ implies that the notion of set has once again a great mathematical significance; and that infinity must be conceived as actual.

Therefore, both the notions of set and of actual infinity, which, as we have seen in §§7.2 and 7.3, had been used implicitly within analysis for some time, were eventually made explicit by Cantor, and became part of the hard core of a new mathematical research programme: Cantorian set theory.

The notion of set has to be considered as an element of the hard core of the new MRP, because this was, after all, generated to study some properties of sets. On the other hand, if we consider the Cantorian theory of transfinite numbers as essential to Cantorian set theory, we cannot fail to see that, since the concept of actual infinity is a necessary condition for having Cantorian transfinite cardinal numbers, the concept of actual infinity must perform in Cantorian set theory the rôle of theoretical posit characteristic of what is found in the hard core of a Lakatosian MRP. In other words, believing that

Principle 4.1. It is mathematically meaningful to study sets

and that

Principle 4.2. It is mathematically meaningful to study actually infinite sets

are presuppositions for starting a programme the aim of which is the construction of a theory of Cantorian transfinite numbers. But at this point it is legitimate to ask the following two questions.

First, how do we know that infinite sets, like the set $\mathbb{I}$ of irrational numbers and sets of greater power than $\mathbb{I}$, exist?

Secondly, is the theory of Cantorian transfinite numbers essential to Cantorian set theory?

Cantor answers the first question in the *Grundlagen* by means of **Principle 4.3**

Principle 4.3. If a mathematical concept $\mathcal{C}$ is well-defined,[9] and is consistent in itself and with the other mathematical notions of the theory to which it belongs, then there exist entities x which fall under the concept $\mathcal{C}$.[10]

The answer to the second question above will be given in the next section.

5 The hard core of a new MRP (2)

The construction of a theory of Cantorian transfinite numbers (from now on I shall drop the 'Cantorian' from the epression 'Cantorian transfinite numbers') was, right from the start, at the very heart of Cantor's work in set theory. According to Hallett, once Cantor discovered that there were subsets of the real line which were infinite, and which had a different size (or numerosity) from one another, he formulated the *problem of the continuum*:[11]

> ...how many powers in all are represented in the real line?

Hallett adds that:[12]

> From Cantor's viewpoint in 1878 there were two possible ways of approaching the continuum problem. One was to try to prove (or disprove) directly that any infinite linear point-set is either denumerable or has the power of the continuum. The second was via a solution to the general problem of powers, by defining an arithmetical

---

[9]A concept $\mathcal{C}$ is well-defined if and only if, for any x, we can decide whether or not $\mathcal{C}(x)$ and for any two entities a and b that are $\mathcal{C}$ we can decide whether or not $a = b$.
[10]See [Cantor, 1883], §8, p. 98:

> Mathematics develops in a completely free way, except that its concepts cannot be self-contradictory, and must be in a definite relationship, regulated by definitions, to those concepts which have been already constructed, are consolidated, and available. When, in particular, new numbers are introduced in mathematics, the only obligation which must be respected is that definitions must be provided which ensure their determinedness, and, in some cases, a relation with already given numbers so that it is possible in each case to distinguish them from one another. As soon as a number satisfies all these conditions, it can and, indeed, must be considered as existing and real in mathematics. This is, in my view, the reason (already mentioned in §4) why rational, irrational, and complex numbers must be considered as existing as in the case of the finite positive integers.

[11][Hallett, 1984], p. 2.
[12][Hallett, 1984], Part 1, Introduction, p. 3.

5. THE HARD CORE OF A NEW MRP (2)

scale of infinite size, showing that all sets must be represented in the scale and then discovering at what place the continuum is represented. If it is represented by the power in the second place, the continuum hypothesis must be correct; if not, then it must be incorrect. Cantor tried both lines of attack, and both in different ways involved the transfinite ordinals.

The first line of attack consisted in the attempt to prove the so-called 'Two-Class Theorem':[13]

Theorem 5.1 (Two-Class Theorem). Every infinite subset of $\mathbb{R}$ is either denumerable or has the power of the continuum.

We must first notice that Cantor's approach to the Two-Class Theorem has nothing to do with the idea that $\mathbb{R}$ can be well-ordered. His strategy in attempting to prove the theorem consists, in fact, in the study of properties of point-sets according to which a point-set P can be decomposed into the sum of two sets. Cantor approached this topic by analyzing the properties of derived sets, which he had introduced into his study of the uniqueness conditions for the representation of functions by means of trigonometrical series.

This is the stage in Cantor's thinking at which the transfinite numbers come into consideration. Although at first transfinite numbers occurred as the superscripts of certain derived sets (the derived sets of the second species), and were considered by Cantor as mere labels to individuate derived sets; in [Cantor, 1883] these were presented as legitimate extensions of the number field.

The construction of an arithmetic of transfinite ordinal and cardinal numbers operated a very important generalization of finite-numbers arithmetic, which gave, as one of its by-products a much better understanding of the differences existing between finite and infinite sets, and of the arithmetical operations defined on them. For example, if a set A is finite then its ordinal number will be the same as its cardinal number; but if $A = \{0, 2, 4, \ldots, 1\}$, and $B = \{1, 3, 5, \ldots, 2, 4\}$, then A and B will have the same cardinal number, but the ordinal number of A will be $\omega + 1$, whereas the ordinal number of B will be $\omega + 2$. Moreover, the commutativity of addition fails when we add, for example, a finite ordinal number n and a transfinite ordinal number α:

$$n + \alpha = \alpha \text{ but } \alpha < \alpha + n.$$

However, in spite of the great wealth of results produced by Cantor in his investigations, his approach to the Two-Class Theorem *via* a descriptive point-set theory was unsuccessful. Cantor's best result being the Cantor-Bendixson theorem for closed subsets of $\mathbb{R}$.[14]

---
[13] See [Moore, 1982], Ch. 1, §1.5, p. 41.
[14]

Cantor's second line of attack on the continuum problem was based on the attempt to prove that, as Moore puts it:

$\mathbb{R}$ has the power of the second number-class,[15] or $2^{\aleph_0} = \aleph_1$.

Cantor's strategy to prove the above result was based on the idea that $\mathbb{R}$ can be well-ordered. By taking ordinal and cardinal numbers to be the same where finite sets are concerned, and defining '... each infinite cardinal as the power of the α-th number class[16] for some ordinal α'[17] he founded the theory of cardinal numbers on the theory of ordinals showing the existence of an infinite scale of cardinal numbers, which begins with the finite cardinal numbers and continues with the transfinite ones or alephs. By means of his well-ordering principle he could then assume that any set can be well-ordered and that, therefore, can be compared with any other set and, in particular, with any set occurring in the scale of cardinal numbers. He then *proved*[18] the aleph-theorem according to which every infinite cardinal is an aleph in the hope, which was much later shown to be vain, that he would succeed in proving that the cardinal number (or power) of $\mathbb{R}$ would be $\aleph_1$.

Both lines of attack on the continuum problem show how central the theory of transfinite numbers is for the development of Cantor's set theory confirming the correctness of thinking that the concept of actual infinity is part of the hard core of the would-be Cantorian set theory MRP. In the first case we can see how the theory of transfinite numbers extends arithmetic beyond the realm of finite numbers, and highlights the difference between finite and infinite in a very sharp way; and in the second case we have the counterintuitive, but extremely productive, foundation of the theory of cardinals on that of ordinals.

Having provided evidence in favour of the idea that some of the elements of the hard core of the would-be Cantor-Zermelo MRP are the concepts of:

Theorem 5.2 (Cantor-Bendixson). :

(a) For any uncountable closed set C of reals, there is a perfect set $P \subseteq C$ such that $C \backslash P$ is at most countable.

(b) Any perfect set of reals has cardinality $2^{\aleph_0}$.

See on this [Dauben, 1980], especially §§5.5–5.7; [Moore, 1982], Ch. 1, §§1.5–1.6, pp. 39–64; [Hallett, 1984], pp. 1–11.

[15] The first number-class is the set of finite ordinals, whereas the second number-class is the set of denumerable ordinals.

[16] For the definition of nth number class see Principle **7.3**.

[17] [Moore, 1982], ch. 1, §1.5, p. 45.

[18] Cantor's proof has been strongly criticized by Zermelo on the basis of the fact that it makes use of '... infinite *successive* arbitrary choices'. Such an idea is flawed, for Zermelo, because 'the intuition of time is applied ... to a process that goes beyond all intuition', in [Hallett, 1984], Ch. 4, §4.1, p. 170. This, incidentally, shows that Cantor did *not* have explicitly available the set-theoretical tool represented by the Axiom of Choice.

set and actual infinity; time has now come to ask whether there are other components of the hard core.

6 The hard core of a new MRP (3)

If the first theory of the Cantor-Zermelo MRP is Cantor's **NST**, in the hard core of the MRP, as this begins with textbfNST, we find, besides principles **4.1**–**4.3**, three other principles isolated by Hallett. These principles, which were aimed by Cantor at providing:[19]

> ...arguments to support the legitimate mathematical employment of actual infinities, especially infinite numbers

are:

Principle 6.1 (Cantor's principle of actual infinity). Any potential infinity presupposes a corresponding actual infinity

Principle 6.2 (Cantor's principle of finitism). The transfinite is on a par with the finite and mathematically is to be treated as far as possible like the finite

Principle 6.3 (Cantor's principle of Absolute infinity). The Absolute infinite cannot be mathematically determined

The function performed by principles **6.1** and **6.2**, whose realist connotation is obvious, is that of promoting the research programme by telling us which research direction to take, that is:

(a) it is legitimate to develop a mathematical theory of the actual infinite, because, as principle **6.1** states, there are such things as infinite completed totalities;

(b) in particular, we are invited to develop an arithmetic involving transfinite numbers, because, according to principle **6.2**, we must treat these entities (the transfinite numbers) as far as possible like finite numbers.

The function of principle **6.3** is, instead, that of establishing limitations to the range of objects of investigation of the MRP. In fact, if, on the one hand, principle **6.3** tells us that there exists such a thing as the Absolute (strong realist thesis), on the other hand, declares off limits the mathematical research direction which has the Absolute as an object of investigation.

Concerning the significance of principle **6.3** in **NST**, Hallett argues that Cantor's conception of the Absolute excludes the idea that his set theory had an unrestricted principle of comprehension.[20] Therefore, according to

---
[19] In [Hallett, 1984], Part 1, Introduction to Part 1, p. 7.
[20] See [Hallett, 1984], pp. 38–39:

Hallett, the introduction in **NST** of principle **6.3** is related to the attempt to eliminate anomalies which, in this particular case, are represented by the set-theoretical paradoxes.

However, in spite of the suggestiveness of Hallett's remarks, a suggestiveness reinforced by what happened in set theory *after* the contradictions appeared, it seems to me that the only rôle that the concept of Absolute plays in Cantor's early thought — say, at least until the publication of [Cantor, 1887–88] — is simply that of distinguishing between increasable and non-increasable actual infinity, providing a justification for the fact that only the first type of actual infinity is a legitimate object of mathematical investigation.

But what did Cantor mean by 'increasable and non-increasable actual infinity'? Cantor clarifies this point in a letter to A. Eulenberg of 28 February 1886 in which he says that[21]

> ... we are here obliged to make a fundamental distinction as we differentiate between:
>
> II^a Increasable Actual Infinity or *Transfinitum*.
> II^b Non-increasable Actual Infinity or *Absolutum*.
>
> The three examples of actual infinity previously mentioned [two of which are the positive integers, and the points lying on a circle] belong all to the class II^a of the Transfinite. In the same way belongs to this class the smallest supra-finite ordinal number, which I call ω; then this can be augmented or increased to the next larger ordinal number $\omega + 1$, this again to $\omega + 2$ and so on. But even the smallest actual-infinite power or cardinal number is a transfinite, and the same holds of the next larger cardinal number and so on.
>
> The Transfinite with its wealth of formations and forms points necessarily at an *Absolute*, at the 'true infinity', whose size can in no way be added to or decreased and which therefore, as to quantity, is to be considered as the absolute maximum. The latter so to speak goes beyond the human power of comprehension and eludes in particular mathematical determination; whereas the transfinite not only fills the wide domain of possibilities concerning the knowledge of God, but also offers a rich and always growing field of ideal research ...

and asserts at the end of the letter that

> ... my theory [of transfinite numbers] is completely different from that of Fontenelle and *is free from any contradiction*. [The italics are mine.]

... Cantor's doctrine of the Absolute explicitly denies that every collection can be a set. And one can argue at least that the universal collection and the collection of all ordinals cannot be Cantorian sets, and this even before there was any suggestion that they are contradictory. Cantor *is* quite vague about what collections should be sets, but when one takes his remarks about the transfinite seriously there is good reason to see in them something more like the iterative universe of the modern axiomatic theory than the indiscriminate 'blanket' universe of naïve set theory.

[21][Cantor, 28.2.1886], pp. 405–406, the translation from the original is mine.

6. THE HARD CORE OF A NEW MRP (3)

Indeed, even independently of the quotation above, there seems to be little evidence in support of the idea that the concept of Absolute was introduced by Cantor as a kind of precaution against the contradictions. On the contrary, there are several reasons which make one think that this was not the case. I shall here list some of these.

First, Cantor's doctrine of the Absolute dates from 1883,[22] whereas the paradoxes started appearing in 1897;[23] and what this suggests is that the doctrine of the Absolute, rather than representing a safe-guard against possible contradictions, was introduced by Cantor to fulfil a very different task. Such a view is confirmed by the fact that Cantor in his [Cantor, 1887–88] asserts — in contrast with Hallett's suppositions concerning his inchoate thoughts about differences existing between sets and other types of collections — that the totality of all cardinal numbers forms a well-ordered set (*Menge*):[24]

> We will see later that the totality of all *cardinal numbers or powers* (the finite and the supra-finite) also form a *well-ordered* set if one imagines them ordered according to size.

It is only in his letter to Dedekind of July 1899 that he writes that the system Taw of all transfinite cardinal numbers cannot be conceived as a set (*Menge*), because of its inconsistency.[25]

What this shows is that Cantor became aware of the paradoxes and, in particular, of the inconsistency of the system of all transfinite cardinal numbers, only years after the introduction of the notion of the Absolute.

Secondly, concerning the relationship between the Cantorian Absolute and what Cantor came to call in his later period 'absolutely infinite multiplicities', it must be pointed out that the latter *do not exist as units*, for

---

[22] See [Cantor, 1883].
[23] See [Burali-Forti, 1897].
[24] [Cantor, 1887–88] in [Hallett, 1984], Ch. 2, §2.2, p. 65.
[25] [Cantor, 28.7.1899] in [Cantor, 1962], Part IV, Appendix, p. 447:

> At this point is raised the question whether or not *all transfinite cardinal numbers* are contained in the system Taw. In other words, is there a [transfinite] set whose power is *not* an Alef?
>
> This question is answered in the negative and the reason for this is that we recognized the inconsistency of the systems Ω and Taw.

The tight relationship present in Cantor's later thought between inconsistency and non-being-a-set is confirmed by another letter sent by Cantor to Dedekind in which he says that systems like that of *all thinkable classes* are inconsistent systems and, therefore, they are not sets, [Cantor, 31.8.1899], p. 448:

> The system T, so also the system S are *not* thus *sets*. Consequently, determined multiplicities are not conceivable as units, that is, it is impossible to conceive such multiplicities as actual 'collections of all their elements'. These are the multiplicities which I call 'inconsistent systems', I call 'sets' the others.

him, because they are inconsistent. But, of course, this cannot apply to the Absolute which is seen by Cantor as 'the actual infinity in God'.

Furthermore, if we consider Cantor's definition of set:[26]

> By an "aggregate" (*Menge*) we are to understand any collection into a whole (*Zusammenfassung zu einem Ganzen*) M of definite and separate objects m of our intuition or our thought. These objects are called the "elements" of M.

we realise that[27]

> According to that "definition" every collection of elements is a set; therefore, for every rule or process by means of which a collection of elements is obtained there is a set which contains exactly the elements which conform to the rule, or are obtained in the process, respectively. The simplest general axiom in this direction is the following
>
> **AXIOM OF COMPREHENSION**: For any condition $\mathcal{C}(x)$ on x there exists a set which contains exactly those elements x which fulfil this condition.

And what this means is that in **NST** there is no diagnosis of what the causes of the inconsistencies are; and it is only such a diagnosis what can lead to a definition of set, and to a consequent restriction on the principle of comprehension, which can avoid the known paradoxes.

Some authors, Hallett in particular, argue that Cantor's later thought concerning absolutely infinite collections is a precursor of the limitation of size principle according to which Cantor's absolutely infinite collections are too large to be sets.

This position is, again, implausible in view of the fact that, for the mathematicians who uphold the limitation of size principle, absolutely infinite collections can be treated as units (proper classes), even though these are not considered to be sets;[28] whereas, for Cantor, such collections *cannot* be treated as units at all.

Before bringing this section to a close, we ought to consider that, in the discussions of the last three sections, several important components of the hard core of the Cantor-Zermelo MRP have been individuated; and that these components are very different in kind from those belonging to the hard core of a typical Lakatosian SRP.

---

[26] [Cantor, 1895–97], §1, p. 85.

[27] [Fraenkel *et alii*, 1973], §3.1, pp. 30-31.

[28] Indeed for those who adopt the axiomatic system **VNB**, what is at the root of the contradictions is not the existence of proper classes, e.g., the class of all sets or that of all ordinals, but the assumption that proper classes can be elements of classes and, in particular, that they can be elements of themselves. Such a diagnosis of the causes of the contradictions leads naturally to the following definition of set:

Definition 6.1. x is a set if there exists a class y such that $x \in y$.

See on this [Drake & Sing, 1996], Ch. 9, pp. 193–197; and [Fraenkel *et alii*, 1973], Ch. II, §7, pp. 119-137.

7. THE CANTOR-ZERMELO MRP: THE PROTECTIVE BELT

In fact, whereas, for Lakatos, the elements of the hard core of an SRP are genuine propositions of the theory, e.g., in the case of the Newtonian mechanics SRP the elements of the hard core are the three laws of motion and the law of gravitation, the elements of what I call 'the hard core of the Cantor-Zermelo MRP' are metaphysical assumptions concerning sets and their existence, the transfinite and the Absolute, etc., i.e., what, with T. S. Kuhn, I am going to call 'metaphysical paradigms'.

Now, it seems to me that such a difference between my view of the content of the hard core of an MRP and Lakatos's ideas about the nature of the elements of the hard core of an SRP might have been caused, among other things, by the fact that Lakatos never applied the methodology of SRPs to mathematics.

In any case, what is crucial here is that the metaphysical nature of the principles belonging to the hard core of an MRP, far from removing the vital distinction between the hard core and the protective belt of an MRP, provides an *internalist* justification for it, that is, a justification which, rather than relying on convention — the arbitrary choice of the members of the mathematical community — appeals to the nature of the statements belonging to the hard core of an MRP.

7 The protective belt of the Cantor-Zermelo MRP

Although the hard core of a Mathematical Research Programme stays the same through the change of theories taking place within the MRP, the situation is very different with regard to the elements of the protective belt of the MRP.

As we shall see in this chapter, auxiliary hypotheses change very frequently in coincidence with the change of theories belonging to a given MRP. It makes, therefore, sense to talk about the auxiliary hypotheses of the Cantor-Zermelo MRP which come about with the advent of **NST**. I shall here mention some of these.

If, on the one hand, the implicit presence of an unrestricted principle of comprehension (see §6) is what guarantees set existence in **NST**, it is, on the other hand, legitimate to ask which are the other auxiliary hypotheses/principles which determine what must we mean by 'set' and which allow the generation of sets from other sets.

From Cantor's definition of set given in the previous section emerges the so-called 'blanket notion' of set. In other words, according to Cantor, by 'set' we must mean any collection of objects of our intuition or of our thought which can be considered as a whole and for which we have *identity conditions* (this is what I take Cantor to mean when he says that the objects of our intuition or of our thought which are the elements of a set must be 'definite and separate').

Indeed, apart from an indirect and very general appeal to consistency — the objects of our intuition or of our thought must be able to form a whole, they must be compossible — Cantor's definition of set appeals neither to properties of collections, e.g., size, self-membership, etc.; nor to the way collections are generated. And it seems to exclude explicitly from sethood only those collections which have vague objects as elements.

Concerning the auxiliary hypotheses which determine a development of the *positive heuristic* of the Cantor-Zermelo MRP by providing principles for the generation of sets from other sets, I am here going to list some of these principles operating a distinction between those defined on any set, and those defined on ordinals.

However, before giving the list of the above mentioned principles, it is important to keep in mind that Cantor holds a realist view of sets. What I mean by this is that Cantor's introduction of principles whereby it is possible to obtain sets from given sets has no constructivist connotation whatsoever. Therefore, in what follows expressions like 'principles of construction', 'principles of generation', etc., when used with regard to Cantor's **NST**, must be understood *realistically*, that is, as mere assertions of existence.

As operations defined on sets in general Cantor had:

$$\bigcap, \bigcup, \times, \mathcal{P}(x), x \backslash y.$$

However, ordinals were originally (in [Cantor, 1883]) generated according to three *inductive* principles:

Principle 7.1. Given an ordinal α, you can generate its successor by adding to it the ordinal number 1;

Principle 7.2. If you have a monotonically increasing sequence of ordinals (α_i) then this sequence identifies a least ordinal β such that, for any $\alpha_i \in (\alpha_i)$, $\alpha_i < \beta$;

Principle 7.3. If we call '1st number class' the set $NC(1)$ of all finite ordinals, we define the nth-number class, for $n > 1$, as the set $NC(n)$ of ordinals α which we can form starting from 1 using Principles **7.1** and **7.2**, and such that, for any $\alpha \in NC(n)$, the cardinal number of $\{\beta : \beta < \alpha\}$ is the same as the cardinal number of $NC(n-1)$.

Principle **7.1**, which Cantor called 'the first principle of production', allows the construction of the first number class, i.e., of the class of ordinals represented by the monotonically increasing sequence (i)

(i) $1, 2, \ldots$ **1st NC**

7. THE CANTOR-ZERMELO MRP: THE PROTECTIVE BELT

Although the first number class does not have a largest element is, nevertheless, completely determined by a specific law of generation (Principle **7.1**) and bounded from above. Since (i) is a monotonically increasing sequence of ordinals, we can apply to it Principle **7.2** which postulates the existence of a least upper bound for the elements of the First Number Class, least upper bound to which Cantor refers using the symbol ω. Therefore, the sequence of ordinals generated so far is monotonically increasing and has the following shape:

$$(ii) \quad 1, 2, \ldots, \omega.$$

If we now apply to (ii) principles **7.7.1** and **7.7.2**, we will extend (ii) into the monotonically increasing sequence of ordinals (iii):

$$(iii) \quad 1, 2, \ldots, \omega, \omega+1, \ldots, \omega 2, \omega 2+1, \ldots, \omega^2, \omega^2+1, \ldots, \omega^\omega, \ldots$$

The function of Principle **7.3** becomes evident at this point and consists in individuating the Second Number Class — separating it out from the First Number Class — as the monotonically increasing sequence of ordinals (iv):

$$(iv) \quad \omega, \omega+1, \ldots, \omega 2, \omega 2+1, \ldots, \omega^2, \omega^2+1, \ldots, \omega^\omega, \ldots \quad \textbf{2nd NC}$$

Although the Second Number Class, like the First Number Class, does not have a largest element, it is completely determined by two specific laws of generation (Principles **7.1** and **7.2**), has a first element and is bounded from above. Since (iv) is a monotonically increasing sequence of ordinals, we can apply to it Principle **7.2** generating the least upper bound Ω of the elements of the Second Number Class. Therefore the ordinals so far generated are:

$$(v) \quad 1, 2, \ldots, \omega, \omega+1, \ldots, \omega 2, \omega 2+1, \ldots, \omega^2, \omega^2+1, \ldots, \omega^\omega, \ldots, \Omega.$$

If we now apply Principles **7.1**, **7.2** and then Principle **7.3** to (v), we will generate the Third Number Class, etc.

However, we must notice that, if we follow the procedures specified above, we will be able to generate only the α-Number Classes of ordinals for $\alpha < \omega$. If we want to generate the ω-Number Class and, in more general terms, α-Number Classes where α is a limit ordinal, we have to replace Principle **7.4** for Principle **7.3**:

Principle 7.4. If we call '1st number class' the set $NC(1)$ of all finite ordinals, we define the γth-Number Class, for $\gamma > 1$, as the set $NC(\gamma)$ of ordinals α which we can form starting from 1 using Principles **7.1** and **7.2**, and such that,

- if γ is NOT a limit ordinal, for any $\alpha \in NC(\gamma)$, the cardinal number of $\{\beta : \beta < \alpha\}$ is the same as the cardinal number of $NC(\gamma - 1)$; and

- if γ IS a limit ordinal, for any $\alpha \in NC(\gamma)$, the cardinal number of $\{\beta : \beta < \alpha\}$ is the same as the cardinal number of

$$\bigcup_{\delta < \gamma} NC(\delta).$$

The above mentioned principles of set formation and the principle of comprehension generate, together with the other auxiliary hypotheses I discussed in the first part of this section, the initial positive heuristic of the Cantor-Zermelo Mathematical Research Programme as this begins to move its first steps with **NST**. In what follows in this study of the evolution of the Cantor-Zermelo set theory MRP, we shall see, among other things, that, indeed, some of the auxiliary hypotheses of **NST** are refuted and replaced by others.

8 Zermelo's system Z: the hard core

At the end of Cantor's activity the system **NST** suffered from a large number of serious problems. On the one hand, results crucial for the development of the theory were still unproved, some such being the well-ordering principle, the trichotomy of cardinals, the aleph-theorem,[29] and the Continuum Hypothesis. And, on the other hand, **NST** had been beset by a number of anomalies/contradictions: the Burali-Forti paradox, Cantor's paradox, Russell's paradox, etc.

As is well known, the situation changed only when Zermelo proved the well-ordering principle and, sometime after, produced the axiomatic system **Z**, which was aimed at showing that his proof of the well-ordering principle was sound and at avoiding the paradoxes which affected **NST**.

It is particularly significant that the proof of the well-ordering principle, and those of the trichotomy of cardinals and of the aleph-theorem, were eventually obtained using the Axiom of Choice (**AC**), because this is a set-theoretical principle — provably equivalent to the well-ordering principle — which is not available in **NST**.

---
[29]

Theorem 8.1 (Well-ordering). Every set can be well-ordered.

Theorem 8.2 (Trichotomy of cardinals). For any cardinal numbers α and β, we have that
$$\alpha < \beta \text{ or } \alpha = \beta \text{ or } \alpha > \beta.$$

Theorem 8.3 (Aleph-theorem). Every transfinite cardinal is an aleph.

8. ZERMELO'S SYSTEM Z: THE HARD CORE

In this and in the following section, I shall briefly discuss Zermelo's system **Z** and argue that **Z** cannot be considered just as a formalization of **NST**, but is a different theory altogether. **Z** is, in fact, a theory which, although sharing the same hard core as **NST**, nevertheless, *falsifies* it.

In **Z** Zermelo assumes the existence of a domain/universe of objects $\mathfrak{B}$ among which are sets, and conceives of the relation '∈' as holding only among some of the objects contained in $\mathfrak{B}$. Zermelo's view of the universe of sets is very important for us, because it reveals that $\mathfrak{B}$ is the counterpart in **Z** of the Cantorian Absolute. Indeed, according to Zermelo, $\mathfrak{B}$ is *not* a set and no mathematics can be done with it. Like the Cantorian Absolute, $\mathfrak{B}$ is a case of non-increasable actual infinity in the sense specified in §6. Such a position on the universe of sets is particularly important, because it shows that Principle **6.3** is what dictates the *negative heuristic* of **Z** represented by the Zermelian doctrine of *limitation of size*.

According to this doctrine:[30]

A collection of sets is a set if and only if it is not equinumerous to the collection of all sets.

Therefore, a collection of sets is a set if its size (numerosity or power) is smaller than that of the collection of all sets.

What motivates the doctrine of limitation of size is the idea that the blame for the paradoxes affecting **NST** must be put squarely on the attempt to do mathematics with collections of sets which are 'too large', where a collection of sets A is too large if A is equinumerous to $\mathfrak{B}$, i.e., if there is a one-one function $f : A \to \mathfrak{B}$ such that $f(A) = \mathfrak{B}$.

One of the consequences of the implementation of the limitation of size doctrine in **Z** is that the meaning of the word 'set' in **Z** differs from the meaning of the word 'set' in **NST**. Therefore, when **Z** takes over from **NST**, one of the things that are eliminated is the **NST** definition of set. This is an extremely important event also from a philosophical point of view, because it shows that a *Gestalt*-shift takes place in going from **NST** to **Z**. This is a *Gestalt*-shift which points out that **NST** and **Z** express very different views about sets, and that the growth in mathematical knowledge brought about by the substitution of **Z** for **NST** is not the outcome of a cumulative process.

I must here emphasize that Zermelo's concept of set in **Z** is not the product of a Koetsier-type operation of conceptual refinement taking place on Cantor's definition of set whereby, as a consequence of being confronted with a large number of exceptions, Cantor's definition is eventually brought to relate only to a smaller range of objects than it was originally supposed

---
[30][Fraenkel et alii, 1973], Chapter II, §5.3, p. 95.

to relate to and is replaced by a more general definition of set. In actual fact, what happened in set theory with the introduction of the **Z**-concept of set was quite the opposite of the process described above. For, Cantor's definition was rejected because it leads to inconsistency and Zermelo introduced a new definition of set which individuates *only* certain types of collections as sets.

But, if we now shift our attention to elements of the hard core of **NST** which differ from Principle **6.3**, it can be easily seen that Principles **6.1** and **6.2** are clearly at the root of the *positive heuristic* of **Z**. In fact, on the one hand, the explicit endorsement of the actual infinite in **Z**, as stated by the Axiom of Infinity (**9.7**), and Zermelo's idea that the universe of sets is given as a totality, must have as necessary condition the acceptance of Principle **6.1** according to which *any potential infinity presupposes a corresponding actual infinity*. And, on the other hand, it is also clear that Principle **6.2** — *the transfinite is on a par with the finite and mathematically is to be treated as far as possible like the finite* — inspires and directs the *positive heuristic* of **Z**. Indeed, Axioms **9.1**-**9.6** of **Z** can be applied to finite and infinite sets alike, and make possible the development of the theory of transfinite numbers within **Z**.

Having shown that the hard core of **Z** is the same as that of **NST**, even though the meaning of the concept of set changes in going from **NST** to **Z**, it is now time to have a closer look at the protective belt of **Z**.

9 The protective belt of Z

As I have said in the previous section, the system **Z** originates, on the one hand, from the desire to justify the successful attempt to solve one of the main problems of the *positive heuristic* of **NST**: the proof of the well-ordering principle; and, on the other hand, from the wish to avoid the paradoxes.[31] And if we examine the axioms of **Z** with some care, we will realize that, as a consequence of the attempt to achieve the above mentioned aims, several important things have changed with respect to the protective belt of **NST**.

First of all, the assertions of existence expressed by axioms **9.2**–**9.7** (see below) are such that: (i) the unrefuted content of **NST**, i.e., what you can prove in **NST** without using methods leading to known paradoxes, is also provable within **Z**; (ii) there are many results provable in **Z** which are not part of the unrefuted content of **NST**. (What this shows is that **Z** satisfies the conditions of *falsifiability* (1) and (2) in relation to **NST**.

---

[31] Just a point of historical interest. One of the main sources of inspiration for Zermelo in the formulation of **Z** was the contribution given by Dedekind to **NST**. See on this [Dedekind, 1963b], and the very illuminating [Gillies, 1982], Ch. 8, pp. 50–58.

9. THE PROTECTIVE BELT OF Z

Secondly, if it is clear that axioms **9.1-9.7** (see below) are assertions which have been formulated having in mind problems belonging to the positive heuristic of **Z**, it is important to show that, with the exception of axiom **9.1** whose aim is to establish identity conditions for sets,[32] the other axioms have been cast taking also into account the dictates of the doctrine of limitation of size, which is the main strategy of defense against the paradoxes adopted by Zermelo.

To see this consider that:

(α) :

Axiom 9.2 (Elementary sets). There is a set with no elements, called the empty set, and for any objects a and b of $\mathfrak{B}$, there exist sets $\{a\}$, $\{b\}$ and $\{a,b\}$.

(α) is an assertion of existence about sets of very small size. To be precise, Axiom **9.2** asserts the existence of sets of size 0: the size of $\emptyset$; and, given sets a and b, of sets of size 1 and 2: $\{a\}, \{b\}, \{a,b\}$.

(β) : Given a set S,

Axiom 9.3 (Schema of Separation). If a propositional function $P(x)$ is *definit* for a set S, then there is a set T containing precisely those elements x of S for which $P(x)$ is true. (For Zermelo, a propositional function $P(x)$ was *definit* for a set S if the membership relation on $\mathfrak{B}$ and the laws of logic determined whether $P(x)$ held for each x in S.)

(β) asserts the existence of a set T which has a size smaller than or equal to that of a given set S. The limitation of size operated by the assertion of existence expressed by axiom schema **9.3** is particularly important, because it causes a number of anomalies that affect **NST** to disappear. In particular the Axiom Schema of Separation defuses Russell's paradox, since, given a set S and the condition $x \notin x$ no contradiction follows. In fact, the only thing that we can prove is that the set

$$T = \{x : x \in S \text{ and } x \notin x\}$$

is not an element of S, that is, we can prove that: if S is a set there exists at least one set, T, which does not belong to it. An immediate

---

[32] **Axiom 9.1 (Extensionality).** If, for the sets S and T, $S \subset T$ and $T \subset S$, then $T = S$; that is, every set is determined by its members.

consequence of this result is that the collection of all sets $\mathfrak{B}$ is not a set and that, therefore, it is not possible to construct Cantor's paradox either.

Moreover, since given the set $[0, 1] \subseteq \mathbb{R}$ and the propositional function $P(x)$: 'x is a real number between 0 and 1 which can be uniquely characterized by sequences of English words of any finite length', we have the formation of paradoxes like Richard's,[33] it follows that the membership relation on $\mathfrak{B}$ and the laws of logic do not determine whether $P(x)$ holds for each x in $[0,1]$, and that, consequently, $P(x)$ is not *definit*. From this follows that **Z**, in contrast with **NST**, is immune to paradoxes like Richard's.

(γ) If we consider

Axiom 9.4 (Power set). If S is a set, then the power set of S is a set.

we realise that given a set of size k, i.e., $|S| = k$, axiom **9.4** asserts the existence of a set $\mathcal{P}(S)$ the size of which is 2^k. In other words, the size of $\mathcal{P}(S)$ is dependent on, and therefore *limited by*, the size of S.

(δ) Given a set S, the

Axiom 9.5 (Union). If S is a set, then the union of S is a set.

---

[33] The best account of Richard's paradox I know is the following. The passage is taken from [Fraenkel *et alii*, 1973], Chapter I, §3.1, pp. 8-9:

> Let us consider all those real numbers between 0 and 1 that can be uniquely characterized by sequences of English words of any finite (but unbounded) length, e.g., 'point eight', 'the positive square root of point zero seven four', 'the smallest number satisfying the condition that the sum of the square of this number and its product by point one equals point three'. Clearly there are only denumerably many such numbers. Let R be their set. R can then be enumerated. Consider any such enumeration. We now characterize a real number r as *that real number between 0 and 1 whose n-th digit after the decimal point is the cyclic sequent of the n-th digit of the n-th number in the enumeration under consideration* (where '1' is the cyclic sequent of '0', ..., and '0' the cyclic sequent of '9'). From [the consideration that the first digit after the decimal point in r differs from the first digit after the decimal point of the first number in the enumeration r_1, it follows that $r \neq r_1$. From the consideration that the second digit after the decimal point in r differs from the second digit after the decimal point of the second number in the enumeration r_2, it follows that $r \neq r_2$. etc. From what has just been argued], it follows that r is different from all the members of R and is therefore not uniquely characterizable by a finite sequence of English words, in plain contradiction to the fact that r has just been characterized in this fashion, viz. by the italicized sequence of English words in the preceding sentence.

9. THE PROTECTIVE BELT OF Z

postulates the existence of the set

$$\bigcup S = \{x : x \in y \text{ where } y \in S\}.$$

The limited size of $\bigcup S$ is shown by the fact that if $\mathcal{F}$ is a collection of finite and denumerable sets such that, for any $S \in \mathcal{F}$, if $x \in S$ then $x \in \mathcal{F}$, then for any $S \in \mathcal{F}$, $\bigcup S$ is either finite or denumerable. If $\mathcal{F}'$ is a collection of sets which are finite, denumerable or have the power of the continuum such that, for any $S \in \mathcal{F}'$, if $x \in S$ then $x \in \mathcal{F}'$, then for any $S \in \mathcal{F}'$, $\bigcup S$ is either finite or denumerable or has the power of the continuum, etc.

(ε) Concerning

Axiom 9.6 (Choice). If S is a disjoint set of non-empty sets, then there is a subset T of the union of S which has exactly one element in common with each member of S.

we must notice that since, given a disjoint set S of non-empty sets, the choice-set T is a subset of $\bigcup S$, it follows that the size of T is limited by the size of $\bigcup S$, which is limited according to the argument given in (δ).

(ζ) Finally, for the

Axiom 9.7 (Infinity). There is a set Z containing the empty set and such that for any object a, if $a \in Z$, then $\{a\} \in Z$.

to be true, it is sufficient that there exist sets which are as small as

$$Z = \{\emptyset, \{\emptyset\}, \{\{\emptyset\}\}, \ldots\}.$$

Having seen the doctrine of limitation of size at work in the formulation of all the axioms of **Z** which consist of assertions of existence, another important thing we must notice with regard to the system **Z** is that this has a model in the cumulative hierarchy of sets V known as the von Neumann universe (see fig. 7.1).

The existence of a model of **Z** provides a confirmation of the excess theoretical content of **Z** with respect to **NST** satisfying *falsifiability* condition (3).

Lastly, some of the axioms of **Z** — the Axiom of Choice — have found surprising applications to mathematical problems completely unrelated to those for the solution of which they were introduced. For instance, 'within' set theory itself with the Axiom of Choice it is possible to prove that $\times_{i \in I} A_i$

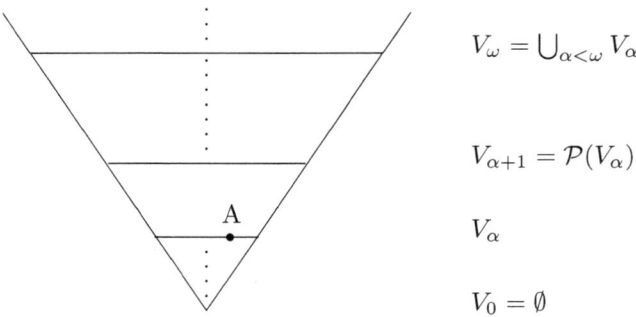

Figure 7.1. The von Neumann Universe $V = \bigcup_{\alpha \in \mathrm{On}} V_\alpha$

and $\bigcup_{i \in I} A_i$ are well-defined, for any infinite set I; $\sum_{i \in \omega} |a_i| = \aleph_0$, for any countable set a_i; the Zermelo-König theorem, etc. And 'outside' set theory it is possible to prove Tychonoff's theorem; that every vector space has a basis; the maximal ideal theorem for lattices, etc. This shows that **Z** satisfies also Zahar's criterion for the *falsifiability* of theories (condition (4)).

From all these considerations we can conclude that **Z** *falsifies* **NST** in Lakatos's sense and that, therefore, the stage of development of the Cantor-Zermelo MRP in going from **NST** to **Z** shows that the MRP is so far *progressive*.

10 Problems with Z, and the system ZFC

Although, as we have just seen, **Z** is a *better theory* than **NST**, **Z** has, nevertheless, become object of criticism. I shall here consider three points.

First, the concept of *definit* propositional function used in **Z** to prevent the formation of Richard-like paradoxes is vague, because:[34]

> [Zermelo's] definition of "definit" property ... invokes "the universally valid laws of logic"; [and] since Zermelo pays no attention at all to the underlying logic, these laws are left unspecified, and the notion of definite property remains hazy.

Secondly, the set-construction principles to be found in some of the axioms of **Z** are not strong enough to enable us to prove the existence of either the set
$$\{\mathbb{N}, \mathcal{P}(\mathbb{N}), \mathcal{P}(\mathcal{P}(\mathbb{N})), \ldots\}$$

---

[34][van Heijenoort, 1967], p. 199.

10. PROBLEMS WITH Z, AND THE SYSTEM ZFC

or of the union of its elements. Therefore, in **Z** we cannot prove the existence of sets which have cardinality α, for $\aleph_\omega \leq \alpha$.

Thirdly, as Zermelo himself admits in talking about **Z**, '... the possibility that $x \in x$ is not in itself excluded by our axioms';[35] and, as we know, the condition $x \in x$ is one of the ingredients to be found in the noxious mixture at the root of several set-theoretic paradoxes.

As a response to the above criticism of **Z**, mathematicians have, over a period of time characterized by much controversy, devised the system of set theory known as **ZFC** which, holding on to the hard core of the Cantor-Zermelo MRP, operates some changes in the protective belt of **Z** which greatly enhance both the *positive* and the *negative heuristics* of **ZFC** with respect to those of **Z**.

First of all, the logic underlining the language of **ZFC** is first-order logic. This choice gives a sharp delineation of what we take the 'universally valid laws of logic' to be, and in so doing answers the first objection raised against **Z** concerning the vagueness of the notion of *definit* propositional function (or property).

Secondly, since the axioms of **ZFC** are first-order versions of the axioms of **Z** plus the Axiom Schema of Replacement and the Axiom of Foundation (see below), we must notice, in particular, that the introduction of the Axiom Schema of Replacement

Axiom 10.1 (Schema of Replacement). If A is a set and $P(x, y)$ formalizes a property of pairs of sets which defines a function of sets, then there exists a set C which contains as elements all the values of this function acting on the elements of A.

supplies a set-construction principle able to show that

$$\{\mathbb{N}, \mathcal{P}(\mathbb{N}), \mathcal{P}(\mathcal{P}(\mathbb{N})), \ldots\}$$

is a set.

In fact, consider the diagram below:

$$\begin{array}{ccccc}
 & \mathbb{N}, & \mathcal{P}(\mathbb{N}), & \mathcal{P}(\mathcal{P}(\mathbb{N})), & \ldots \\
g & \uparrow & \uparrow & \uparrow & \uparrow \\
 & \aleph_0, & \aleph_1, & \aleph_2, & \ldots \\
f & \uparrow & \uparrow & \uparrow & \uparrow \\
 & 0, & 1, & 2, & \ldots
\end{array}$$

Since f, g, and $g \circ f$ are one-one functions and $\{0, 1, 2, \ldots\}$ is a set (by the Axiom of Infinity); it follows, by the Axiom Schema of Replacement,

---

[35] [Zermelo, 1908], ibid., p. 203.

that $\{\aleph_0, \aleph_1, \aleph_2, \ldots\}$ and $\{\mathbb{N}, \mathcal{P}(\mathbb{N}), \mathcal{P}(\mathcal{P}(\mathbb{N})), \ldots\}$ are sets. Since

$$\{\mathbb{N}, \mathcal{P}(\mathbb{N}), \mathcal{P}(\mathcal{P}(\mathbb{N})), \ldots\}$$

is a set then, by the Axiom of Union, also $\mathbb{N} \cup \mathcal{P}(\mathbb{N}) \cup \mathcal{P}(\mathcal{P}(\mathbb{N})) \cup \ldots$ is a set.[36]

However, the importance of the Axiom Schema of Replacement is not confined to set theory, but extends to other branches of mathematics:[37]

> Until recently, the important known consequences of Replacement had all been squarely within set theory itself..., but in the mid-1970s, Martin used the axiom to establish a welcome regularity property of certain definable sets of reals (see [Martin, 1975]). A few years earlier, Harvey Friedman had shown that this conclusion could not be proved from Zermelo's theory alone (see [Friedman, 1971]). Thus Martin's theorem demonstrates that Replacement plays a central role in the theory of sets of reals, a theory of central interest to the early analysts.

Therefore, in **ZFC** we can prove the existence and study the properties of sets the existence of which cannot be proved in **Z**, and give contributions to branches of mathematics other than set theory which it is not possible to give within **Z**. This implies that **ZFC** has greater theoretical content than **Z**. (Satisfaction of *falsifiability* criterion (1).) And since some of the contributions given by **ZFC** to set theory and other branches of mathematics are unrelated to the reasons why, for example, the Axiom Schema of Replacement was introduced, we can conclude that **ZFC** satisfies *falsifiability* criterion (4) with respect to **Z**.

Thirdly, as a consequence of the Axiom of Foundation,

Axiom 10.2 (Foundation). If A is a set, then either A is empty (i.e., A has no elements) or there exists an element C of A which is disjoint from A.

it is impossible to have in **ZFC** a set A such that

(1) $\quad A \in A$.

In fact, for any set A, $\{A\}$ is a set (axiom **9.2**) and it is non-empty. Therefore, by the Axiom of Foundation

$$A \cap \{A\} = \emptyset.$$

---

[36]Notice that the Axiom schema of Replacement respects the doctrine of limitation of size, because if S is a set on which is defined a function f, then

$$|S| \geq |f(S)|,$$

even though the Axiom schema of Replacement, taken in combination with other axioms of **ZFC**, can be used to prove the existence of sets whose cardinal numbers are very large.

[37][Maddy, 2000], Ch. 1, §3, p. 59. The result proved by Martin using the Axiom of Replacement is that all Borel sets are determined.

10. PROBLEMS WITH Z, AND THE SYSTEM ZFC

From this follows that $A \notin A$. We cannot even have situations like the following:

$$(2) \quad A \in B_n \in B_{n-1} \in \cdots \in B_1 \in A,$$

where $A, B_1, B_2, \ldots, B_n$ are sets, because by repeated applications of the Axiom of Elementary Sets and of the Axiom of Union it is possible to show that $C = \{A, B_1, \ldots, B_n\}$ is a set; but since C is non-empty and, for any $x \in C$, we have $x \cap C \neq \emptyset$, we would contradict the Axiom of Foundation. Likewise if $A, B_1, \ldots$ are sets, we cannot have

$$(3) \quad \cdots \in B_n \in \cdots \in B_1 \in A,$$

because in this case, using the Axiom Schema of Replacement, it is possible to show that $D = \{A, B_1, B_2, \ldots\}$ is a set, and since D is non-empty and, for any $x \in D$, we have $x \cap D \neq \emptyset$ this would contradict the Axiom of Foundation.

Therefore, since conditions (1)–(3) give rise to Russell-type antinomies, we have that a further source of trouble still present in **Z** is removed in **ZFC** through an extension of the *negative heuristic* of **Z**, an extension obtained with the introduction of the Axiom of Foundation. Such an extension implies that, for a collection of sets A to be a set, A, besides not being too large, must also be well-founded, i.e., there must be an $\alpha \in \mathrm{On}$ such that $A \in V_\alpha$, where V_α is a level of the cumulative hierarchy of sets V. The change of meaning of the word 'set' in going from **Z** to **ZFC** shows once more that the growth of set-theoretical knowledge within the Cantor-Zermelo MRP is not cumulative.

Lastly, the unrefuted content of **Z** is clearly part of **ZFC** (satisfaction of *falsifiability* criterion (2)); V, as well as being a model of **Z**, is a model of **ZFC** (satisfaction of *falsifiability* criterion (3)), and in spite of the fact that **ZFC** is incomplete (Gödel's First Incompleteness Theorem), non-categorical (Löweneheim-Skolem theorem, and Gödel's First Incompleteness Theorem), that its consistency can be proved only in a larger system (Gödel Second Incompleteness Theorem), and that some of its axioms seem to have paradoxical consequences (see §2.4, footnote 12), **ZFC** has come to play a very central rôle in mathematics by providing a unifying language and framework within which to express mathematical theories. This is an important result which, like that concerning the relevance of the Axiom Schema of Replacement to the theory of definable sets of reals, is not connected with the reasons why **ZFC** was formulated in the first place. From this, we can conclude that **ZFC** satisfies Zahar's criterion of *falsifiability* with respect to **Z** and, consequently, that **ZFC** *falsifies* **Z** in Lakatos's sense. From this follows that also the stage of development of the Cantor-Zermelo MRP which sees **ZFC** taking over from **Z** shows that the Cantor-Zermelo MRP has so far been *progressive*.

It is of some comfort to know that the progressive character manifested by the Cantor-Zermelo MRP has remained intact through the years, as it is shown by Cohen's results concerning the independence of **CH** from **ZFC** and of **AC** from **ZF**,[38] and by the large amount of important recent work on large cardinals.

11 Anti-realist rivals of the Cantor-Zermelo MRP

We have seen in this chapter that Cantor-Zermelo set theory is a *progressive* MRP with very strong realist inclinations, a *progressive* MRP which is mainly concerned with the foundations of mathematics.

At this point of the discussion, it is natural to ask whether or not there are *progressive* MRPs in the foundations of mathematics whose hard core contains strong anti-realist assumptions, and which can rival the Cantor-Zermelo MRP.

Someone may try to answer this question in the affirmative suggesting the idea that the application of the axiomatic method to any mathematical theory can be seen as part of an anti-realist MRP whose task is to show that mathematical theories are games.

Now, independently of the need, on the part of the proponent of such a suggestion, to specify the hard core and the protective belt of this formalist MRP, we can argue, as we did in §4.3, that the axiomatic method is neutral with respect to the philosophical bias of those of its supporters who have an anti-realist view of mathematics. For, when we study the history of mathematics, we realize that, in most cases, the process of axiom choice, far from being the result of mere convention, is actually driven by the desire of 'making the content of an informal mathematical theory explicit and clearly laid out'.

Another possible challenger of Cantor-Zermelo set theory is the programme of nominalization of the empirical sciences which is at the root of Field's fictionalism. In fact, it is conceivable that there might be someone prepared to assert that Field's programme concerning the nominalization of the empirical sciences is an MRP whose goal is to show that mathematical statements are not true, if they are taken at face value, etc.

To this suggestion it is possible to reply that, first, Field's programme for the nominalization of the empirical sciences is not a mathematical research programme, because its object of study is represented by theories belonging to the empirical sciences and their possible nominalistic translations/reductions.

---

[38]Notice that **ZF** = **ZFC** − **AC**.

11. ANTI-REALIST RIVALS OF THE CANTOR-ZERMELO MRP 251

Secondly, even if Field's programme for the nominalization of the empirical sciences were conceivable as a Lakatosian MRP, this, given the insurmountable obstacles present in the way of its realization, would certainly not be *progressive* in Lakatos's sense, and, therefore, could not be considered as a serious rival of the Cantor-Zermelo MRP.

A more creditable candidate for the rôle of anti-realist rival of the Cantor-Zermelo MRP than the two we have examined so far is mathematical constructivism. Indeed, the ambition of constructivism is that of producing a (methodological) unification of mathematical theories through: the adoption of constructive proof-procedures in mathematics to the exclusion of non-constructive ones; the implementation of the BHK-interpretation of the logical constants which leads to the rejection of classical logic; the rejection of the actual infinite; and the acceptance of the idea that mathematics is a creative rather than a descriptive activity of the human mind.

If we now pause to reflect on the principles I have just mentioned as characterizing the main features of constructivism, and on what I have been saying in this chapter concerning Cantor-Zermelo set theory, it will immediately become apparent that constructivism is an anti-realist framework for mathematics which is incompatible with Cantor-Zermelo set theory. The next question we need to address is, of course, whether constructivism is conceivable as a Lakatosian MRP.

We know that within constructivism there are a number of fundamental principles — some of which I have mentioned above — which could be considered as belonging to the hard-core of a mathematical research programme.

However, if constructivism has to be considered as a Lakatosian MRP, the hard-core of constructivism must give rise to a positive heuristic. And the embarrassing thing here is that, when we start looking for the positive heuristic of constructivism, what we actually find in doing number theory, mathematical analysis, etc. is a bundle of constructivist practices the elements of which are accepted only by some of the different schools of constructive mathematics we have examined in chapter 4, and not by others.

This situation makes us realize that mathematical constructivism is presently in a pre-MRP stage of development. For, much of the disagreement existing between Intuitionism, Markov's constructivism, Bishop's mathematics, and Strict finitism has to do with some of the elements of what, in the course of time, could become the hard core of a Lakatosian MRP. One such element being the very concept of mathematical construction.

The present predicament of constructivism, of finding itself in a pre-MRP stage of development, receives a further confirmation from the observation that, since Intuitionism, Markov's constructivism, and Strict finitism are extensionally incommensurable[39] with one another, it is impossible for one of these systems to *falsify* the other two, because none of these systems can satisfy *falsifiability* condition (2) with regard to any of the others.

Taking for granted the correctness of the argument above to the effect that mathematical constructivism is presently not conceivable as a Lakatosian MRP, it is possible to wonder whether individual schools of constructivist thought in mathematics are conceivable as Lakatosian MRPs in their own right, and, if this is the case, whether they can rival Cantor-Zermelo set theory. These questions are particularly cogent, if they are directed to Intuitionism, which appears to be the most philosophically sophisticated and mathematically interesting form of mathematical constructivism produced so far.

In considering Intuitionism as a possible candidate for the status of Lakatosian MRP, we discover in it the reproduction of the hard core directed *bellum omnium contra omnes* we observed in our study of constructivism.

To see this, consider that one of the most important differences existing between Brouwer's Intuitionism and Dummett's is the attitude these two authors have towards a potential hard core component of an Intuitionistic MRP: language.

Indeed, since for Brouwer, language has only a perfunctory rôle in mathematical activity, a rôle which consists in recording and communicating mathematical constructions, it follows that, for him, neither language nor meaning belong to the hard core of an eventual Intuitionistic MRP.

On the other hand, since, according to Dummett, language has an absolutely central function to perform within mathematical activity — as is shown by the fact that Dummett founds his form of Intuitionism, and the justifications for adopting it, on considerations which relate to the meaning and the truth of mathematical statements — it follows that, for him, meaning and language would have to loom large among the elements of the hard core of an Intuitionistic MRP.

If we restrict even more our considerations to Dummett's Intuitionism, to the exclusion of Brouwer's, we observe that, even if there has been some mathematical development within it at the hands of D. Prawitz and P. Martin-Löf, there are still deep hard core disagreements between Dummett, Prawitz, and Martin-Löf. I shall mention only one of these.

We saw in §1.8 that, for Dummett, we should discard the notion of 'provability' within Intuitionism in the sense that, according to the intuitionist,

---

[39]Two mathematical theories T_1 and T_2 are *extensionally incommensurable* if and only if there exists at least one statement S belonging to the languages of T_1 and T_2 such that S is true in T_1 and false T_2 or *vice versa*.

before a mathematical statement S is proved or refuted, it is not determined whether S is provable. This, we saw, is not the case according to Prawitz, for whom proofs of mathematical statements exist even before we find them.

It is important to notice that the above mentioned difference existing between Dummett's and Prawitz's views on provability affects the hard core of an eventual Dummettian Intuitionistic MRP. For, whereas in Prawitz's case his hard core commitment to the meaningfulness of the notion of provability — a notion which he also places at the root of the concept of truth — forces him to commit himself to the belief in the existence of mathematical proofs independently of the activity of producing them and, therefore, forces him to commit himself to a form of realism about proofs, this is not so in Dummett's case, because he considers the notion of provability to be intuitionistically unacceptable.

12   A new argument for realism

The considerations contained in this chapter make possible the formulation of a new argument in favour of realism in mathematics, an argument with which I intend to bring the book to a close.

Sections 7.4-7.10 have been devoted to showing that Cantor-Zermelo set theory is a *progressive* Lakatosian MRP with strong realist inclinations, a Lakatosian MRP which performs a very special foundational rôle within mathematics in the sense that it provides a unification of mathematical theories.

Now, since an MRP comes as a whole in which the hard core is inseparable from the positive heuristic, it follows that accepting an MRP is an act that commits one to accepting the realist or anti-realist metaphysical assumptions contained in its hard core. Therefore, if we, in particular, accept Cantor-Zermelo set theory, we commit ourselves to a realist view of set theory, and of the mathematical theories unified by it.

At this point it is, of course, of paramount importance to ask which criteria of choice we should adopt to select rationally the best between two rival MRPs. Following closely the strategy adopted by Lakatos to solve a similar problem involving rival research programmes present in the empirical sciences, I am going to suggest that the main criterion to adjudicate between two rival MRPs consists in determining whether each of the rival MRPs is *progressive* or *degenerative*; and that, indeed, if one of the two MRPs is *progressive*, whereas the other is *degenerative*, we should then choose the *progressive* MRP.

To see the reason why the one described above is a rational and compelling criterion of choice between rival MRPs, consider that Lakatos's ultimate motivation for introducing the *progressive/degenerative* distinction between research programmes was to find a demarcation line between scientific and pseudo-scientific research programmes present in the empirical sciences. Such a demarcation line was meant by Lakatos to separate those research programmes which publicly and objectively produce an increase in knowledge in their field of investigation (scientific research programmes) from those which do not (pseudo-scientific research programmes).

For Lakatos:[40]

> We 'accept' problemshifts [research programmes] as 'scientific' only if they are at least theoretically progressive; if they are not, we '*reject*' them as 'pseudo-scientific'.

Independently of the important question concerning whether or not Lakatos's proposal succeeds in providing a demarcation line between science and pseudo-science, if we apply the *progressive/degenerative* distinction to MRPs, we observe that this works as a relevant criterion of choice between rival MRPs.[41]

In fact, if an MRP is *progressive* then, it consists of a finite sequence of mathematical theories

$$T_1, \ldots, T_n$$

produced in response to a given hard core common to all of them, and to a number of anomalies generated as a consequence of developing a positive heuristic.

The theories belonging to the sequence above are such that theory T_{n+1}, for $n \in \mathbb{N}$ and $n \neq 0$, *falsifies* theory T_n, where this means that: (1) T_{n+1} has excess theoretical content with regard to T_n, (2) there is a confirmation of the excess theoretical content of T_{n+1} with respect to T_n, (3) all the unrefuted content of T_n is included in the content of T_{n+1}, and that (4) the excess theoretical content of T_{n+1} with respect to T_n must not be the consequence of the occurrence within T_{n+1} of *ad hoc* hypotheses.

A very important thing we must notice at this point is that the *falsifiability* criteria (1)-(4) are public, objective, and provide a logically compelling justification for saying that, if the sequence $T_1, \ldots, T_n$ satisfies them then there clearly is a genuine growth of mathematical knowledge within the MRP exemplified by it.

However, if an MRP is *degenerative*, there must be at least two theories T_m and T_n — for $m, n \in \mathbb{N}$ and $m \neq n$, — belonging to it such that T_m

---

[40]See [Lakatos, 1983a], §2(c), p. 34.

[41]Obviously, there are many irrelevant criteria of choice between rival MRPs, one such being the (cultural) authority of the supporters of each of the rival MRPs. On the other hand, an application of the *progressive/degenerative* distinction to MRPs is a relevant criterion of choice between rival MRPs, because, independently of considerations relating to psycho-sociological factors involving the mathematical community which operates the choice, it focuses on the ability of the individual MRPs to contribute to the growth of mathematical knowledge in their field of investigation.

12. A NEW ARGUMENT FOR REALISM

does not *falsify* T_n, and *vice versa*. But, of course, if this is the case then in going from T_m to T_n and from T_n to T_m we are not in a position to tell whether or not there is a genuine growth of mathematical knowledge. For, either T_m and T_n have no excess theoretical content with regard to each other — they are essentially the same theory — or one of them has excess theoretical content with regard to the other, but there is no confirmation of this — *via* a model — or not all the unrefuted content of one is present in the other — there is some loss of information in going from one theory to the other and *vice versa* — or the excess theoretical content that one theory has with respect to the other is the consequence of using arbitrary *ad hoc* assumptions.

From these considerations follows that, if an MRP is *degenerative*, we are not in a position to tell whether or not there is a genuine growth of mathematical knowledge within it.

The considerations just made with regard to *degenerative* MRPs apply *a fortiori* to mathematical theories which find themselves in a pre-MRP stage of development. For, also in this case, there is no rational way of assessing the contribution of these theories to the growth of mathematical knowledge.

If we now use the *progressive/degenerative* distinction to compare what I have taken to be the two main rival framework theories in the foundations of mathematics, that is, Cantor-Zermelo set theory and constructivism (or eventual sub-research programmes of constructivism such as Intuitionism), we realize that whereas Cantor-Zermelo set theory — which has in its hard core very strong realist metaphysical assumptions — is a *progressive* MRP, constructivism (or Intuitionism) — which has in its hard core equally strong anti-realist metaphysical assumptions — is a framework theory which is still in a pre-MRP stage of development.

From this follows that, if we were to choose rationally which, between Cantor-Zermelo set theory and constructivism (or Intuitionism) is the best framework theory presently available for mathematics, we should choose Cantor-Zermelo set theory; and, consequently, we should also have a commitment in favour of realism about mathematics.

APPENDIX I

THE BHK-INTERPRETATION

The Brouwer, Heyting, Kolmogorov (**BHK**) interpretation of the logical constants is expressed by the following axioms:

Axiom I..1. A proof of $A \wedge B$ is given by presenting a proof of A and a proof of B.

A proof of $A \wedge B$ is a verification of the claim made by the assertion of $A \wedge B$, namely, that A is provable and B is provable.

Axiom I..2. A proof of $A \vee B$ is given by presenting either a proof of A or a proof of B (plus the stipulation that we want to regard the proof presented as evidence for $A \vee B$).

Axiom I..3. A proof of $A \longrightarrow B$ is a construction which permits us to transform any proof of A into a proof of B.

A proof of $A \to B$ is a verification of the claim made by the assertion of $A \to B$, namely, that if A is provable then also B is provable.

Axiom I..4. Absurdity $\bot$ (contradiction) has no proof; a proof of $\neg A$ is a construction which transforms any hypothetical proof of A into a proof of a contradiction.

A proof of $\neg A$ is a verification of the claim made by the assertion $A \to \bot$, namely, that if A is provable then $\bot$ is provable.

Axiom I..5. A proof of $\forall x A(x)$ is a construction which transforms a proof of $d \in D$ (where D is the intended range of the variable x) into a proof of $A(d)$.

A proof of $\forall x A(x)$ is a verification of the claim made by the assertion $d \in D \to A(d)$, namely, that if it is provable that $d \in D$ then it is provable that $A(d)$.

Axiom I..6. A proof of $\exists x A(x)$ is given by providing a $d \in D$, and a proof of $A(d)$.

A proof of $\exists x A(x)$ is a verification of the claim made by the assertion $d \in D \wedge A(d)$, namely, that we can find a d, prove that $d \in D$ and that $A(d)$.

BIBLIOGRAPHY

[Aristotle, *Metaphysics*] Aristotle: *Metaphysics*, in [Barnes, 1991], **vol. 2**, pp. 1552-1728.

[Aristotele, *Physics*] Aristotele: *Physics*, in [Barnes, 1991], **vol. 1**, pp. 315-446.

[Aristotle, *Posterior Analytics*] Aristotle: 1975, *Posterior Analytics*, in [Barnes, 1991], **vol. 1**, pp. 114-166.

[Aristotle, *Topics*] Aristotle: *Topics*, in [Barnes, 1991], **vol. 1**, pp. 167-277.

[Aspray & Kitcher, 1988] Aspray, W. & Kitcher, P. (eds.): 1988, *History and Philosophy of Modern Mathematics*, University of Minnesota Press, Minnesota.

[Assenza *et alii*, 2004] Assenza, E. & Chiricò, D. & Perconti, P. (eds.): 2004, *Logic, Ontology and Linguistics*, Rubbettino Editore, Soveria Mannelli.

[Avellone *et alii*, 2002] Avellone, M. & Brigaglia, A. & Zappulla, C.: 2002, 'The Foundations of Projective Geometry in Italy from De Paolis to Pieri', in *Archive for History of Exact Sciences*, **vol. 56**, pp. 363–425, Springer-Verlag.

[Balaguer, 1998] Balaguer, M.: 1998, 'Non-Uniqueness as a Non-Problem', *Philosophia Mathematica*, (3), **vol. 6**, pp. 63-84.

[Baldwin, 1993] Baldwin, T. (ed.): 1993, *G. E. Moore: Selected Writings*, Routledge, London.

[Barnes, 1991] Barnes, J. (ed.): 1991, *The complete works of Aristotle*, Bollingen series LXXI, Princeton University Press, Princeton, New Jersey.

[Benacerraf, 1985] Benacerraf, P.: 1985, 'What numbers could not be', in [Benacerraf & Putnam, 1985], pp. 272–294.

[Benacerraf & Putnam, 1985] Benacerraf, P. & Putnam, H. (eds.): 1985, *Philosophy of Mathematics*, Selected Readings, Second Edition, Cambridge University Press, Cambridge.

[Berkeley, 1734] Berkeley, G.: 1734, *The Analyst*, in [Ewald, 1996], **vol. 1**, pp. 60–92.

[Billinge, 2003] Billinge, H.: 2003, 'Did Bishop have a philosophy of mathematics?', *Philosophia Mathematica*, (3), **vol. 11**, pp. 176–194.

[Bishop & Bridges, 1985] Bishop, E & Bridges, D.: 1985, *Constructive Analysis*, Springer-Verlag, New York, Berlin, Heidelberg, Tokio.

[Bonola, 1955] Bonola, R.: 1955, *Non-Euclidean geometry*, transl. by H. S. Carslaw, Dover publications Inc., New York.

[Bourbaki, 1948] Bourbaki, N.: 1948, 'The Architecture of Mathematics', in [Ewald, 1996], **vol. 2**, pp. 1265-1276.

[Boyer, 1949] Boyer, C. B.: 1949, *The history of the calculus and its conceptual development*, Dover Publications Inc., New York.

[Boyer, 1968] Boyer, C.B.: 1968, *A history of mathematics*, John Wiley & sons, Chichester.

[Bridges, 1998] Bridges, D.: 1998, 'Constructive Truth in Practice', in [Dales & Oliveri, 1998], pp. 53–69.

[Bridges & Richman, 1988] Bridges, D. & Richman, F.: 1988, *Varieties of constructive mathematics*, London Mathematical Society Lecture Note Series, Cambridge University Press, Cambridge.

[Brouwer, 1907] Brouwer, L. E. J.: 1907, *On the foundations of mathematics. Dissertation*, in [Brouwer, 1975], pp. 11-101.

[Brouwer, 1912] Brouwer, L.E.J.: 1912, 'Intuitionism and Formalism', [Benacerraf & Putnam, 1985], pp. 77–89.

[Brouwer, 1975] Brouwer, L. E. J.: 1975, *Collected Works I*, ed. by A. Heyring, North-Holland P.C., Amsterdam.

[Brouwer, 1981] Brouwer, L. E. J.: 1981, *Cambridge lectures*, Cambridge University Press, Cambridge.

[Budd, 1987] Budd, M.: 1987, 'Wittgenstein on seeing aspects', *Mind*, **96**, pp. 1-17.

[Burali-Forti, 1897] Burali-Forti, Cesare: 1897, 'Una Questione sui Numeri Transfiniti', *Rendiconti del Circolo Matematico di Palermo*, **vol. 11**.

[Burgess, 2004] Burgess, J.P.: 2004, 'Mathematics and *Bleak House*', *Philosophia Mathematica*, (3), **vol. 12**, pp. 18–36.
[Burgess & Rosen, 1997] Burgess, J.P. & Rosen, G.: 1997, *A Subject With No Object*, Clarendon Press, Oxford.
[Cantor, 1879] Cantor, G.: 1879, 'Über unendliche, lineare Punktmannigfaltgkeiten I', *Mathematische Annalen*, **vol. 15**, pp. 1-7.
[Cantor, 1882] Cantor, G.: 1882, 'Über unendliche, lineare Punktmannigfaltgkeiten III', in [Cantor, 1962], pp. 149-157.
[Cantor, 1883] Cantor, G.: 1883, 'Über unendliche, lineare Punktmannigfaltigkeiten V' — *Grundlagen einer allgemeinen Mannigfaltigkeitslehre* — in [Cantor, 1962], pp. 165-209.
[Cantor, 28.2.1886] Cantor, E.: 1886, 'Cantor an Eulenburg', in [Cantor, 1962], pp. 400–407.
[Cantor, 1887–88] Cantor, G.: 1887-88, 'Mitteilungen zur Lehre vom Transfiniten', in [Cantor, 1962], pp. 378–439.
[Cantor, 1895–97] Cantor, G.: 1895/97 [1955], *Contributions to the Founding of the theory of Transfinite Numbers*, Dover Publications Inc., New York.
[Cantor, 28.7.1899] Cantor, G.: 1899, 'Cantor an Dedekind', in: [Cantor, 1962], pp. 443–447.
[Cantor, 31.8.1899] Cantor, G.: 1899, 'Cantor an Dedekind', in [Cantor, 1962], p. 448.
[Cantor, 1962] Cantor, G.: 1962, *Gesammelte Abhandlungen*, E. Zermelo Editor, Georg Olms Verlagsbuchhandlung, Hildesheim.
[Carnap, 1932] Carnap, R.: 1932, 'Überwindung der Metaphysik durch logische Analyse der Sprache', *Erkenntnis*, **II**, pp. 219–241.
[Carnap, 1963] Carnap, R.: 1963, *W. V. Quine on logical truth*, in [Schilpp, 1963], pp. 915-922.
[Carnap, 1985] Carnap, R.: 1985, 'Empiricism, semantics, and ontology', in: [Benacerraf & Putnam, 1985], pp. 241–257.
[Cellucci, 1998] Cellucci, C.: 1998, *Le ragioni della logica*, Laterza, Roma-Bari.
[Cellucci & Gillies, 2005] Cellucci, C. & Gillies, D. (eds.): 2005, *Mathematical Reasoning and Heuristics*, King's College Publications, London.
[Church, 1951] Church, A.: 1951, 'The need for abstract entities in semantic analysis', *Proceedings of the American Academy of Arts and Sciences*, **vol. 80**, pp. 100-112.
[Corfield 2003,] Corfield, D.: 2003, *Towards a Philosophy of Real Mathematics*, Cambridge University Press, Cambridge.
[Cottingham et alii, 1987] Cottingham, J. & Stoothoff, R. & Murdoch, D. (transl.): 1987, *The Philosophical Writings of Descartes*, Cambridge University Press, Cambridge.
[Curry, 1954] Curry, H. B.: 1954, 'Remarks on the definition and nature of mathematics', in [Benacerraf & Putnam, 1985], pp. 202-206.
[Cutland, 1988] Cutland, N. J.: 1988, *Computability*, Cambridge University Press, Cambridge.
[Dales, 1998] Dales, H.G.: 1998, 'The mathematician as a formalist', in [Dales & Oliveri, 1998], pp. 181–200.
[Dales & Oliveri, 1998] Dales, H.G. & Oliveri, G. (eds.): 1998, *Truth in Mathematics*, Oxford University Press, Oxford.
[Dauben, 1980] Dauben, J. W.: 1980, 'The Development of Cantorian Set Theory', in [Grattan-Guinness, 1980], pp. 181–219.
[Dauben, 1990] Dauben, J. W.: 1990, *Georg Cantor*, Princeton University Press, Princeton, New Jersey.
[Dedekind, 1963a] Dedekind, R.: 1963, *Essays on the Theory of Numbers*, Dover Publications Inc., New York.
[Dedekind, 1963b] Dedekind, R.: 1963, 'The nature and meaning of numbers', in [Dedekind, 1963a], p. 31-115.
[De Ruggiero, 1967] De Ruggiero, G.: 1967, *Storia della Filosofia*, Laterza, Bari.

[Descartes, 1644] Descartes, R.: 1644 (1987), *Principles of Philosophy*, in [Cottingham et alii, 1987], **vol. I**, pp. 177-291.
[Dieudonné, 1968] Dieudonné, J.: 1968, *Algèbre linéaire et géométrie élémentaire*, (3^a, ed.), Hermann, Paris.
[Drake & Sing, 1996] Drake, F.R. & Singh, D., 1996, *Intermediate Set Theory*, John Wiley & sons, Chichester.
[Dummett, 1973] Dummett, M.A.E.: 1973, 'The philosophical basis of intuitionistic logic' in [Benacerraf & Putnam, 1985], pp. 97-129.
[Dummett, 1978] Dummett, M.A.E.: 1978, *Truth and Other Enigmas*, Harvard University Press, Cambridge, Massachusetts.
[Dummett, 1987] Dummett, M. A. E.: 1987, 'Reply to Prawitz', in [Taylor, 1987], pp. 281-286.
[Dummett, 1991] Dummett, M. A. E.: 1991, *The Logical Basis of Metaphysics*, Duckworth, London.
[Dummett, 1994] Dummett, M. A. E.: 1994, 'Reply to Oliveri', in [McGuinness & Oliveri, 1994], pp. 299-307.
[Dummett, 1995] Dummett, M.A.E.: 1995, *Frege. Philosophy of Mathematics*, Duckworth, London.
[Dummett, 1998] Dummett, M. A. E.: 1998, 'Neo-Fregeans: In Bad Company?', in [Schirn, 1998], pp. 369–87.
[Dummett, 2000] Dummett, M.: 2000, *Elements of Intuitionism*, Second Edition, Clarendon Press, Oxford.
[Ebbinghaus et alii, 1991] Ebbinghaus, H.-D. et alii: 1991, *Numbers*, Springer-Verlag, GTM, New York, Berlin, etc.
[Ernest, 1997] Ernest, P.: 1997, 'The Legacy of Lakatos: Reconceptualising the Philosophy of Mathematics', *Philosophia Mathematica*, (3), **vol. 5**, pp. 116–134.
[Ernest, 1999] Ernest, P.: 1999, 'Review of G. Oliveri, "Mathematics. A science of Patterns?" ', *MathSciNet, Mathematical Reviews on the Web*, American Mathematical Society, **99g:00023** 00A30 (03A05).
[Euclid, The Elements] Euclid: 1956, *The thirteen Books of the Elements*, translated with introduction and commentary by Sir Thomas L. Heath, Second Edition Unabridged, translated from the text of Heiberg, Dover Publications Inc., New York.
[Ewald, 1996] Ewald, W. B. (ed.): 1996, *From Kant to Hilbert*, Clarendon Press, Oxford.
[Field, 1980] Field, H.: 1980, *Science without numbers: A defense of Nominalism*, Basil Blackwell Publisher, Oxford.
[Field, 1989] Field, H.: 1989, *Realism, Mathematics, and Modality*, B. Blackwell, Oxford.
[Field, 1989a] Field, H.: 1989, 'Can we dispense with space-time?', in [Field, 1989], pp. 171–226.
[Field, 2001] Field, H.: 2001, *Truth and the Absence of Fact*, Clarendon Press, Oxford.
[Føllesdal 1999,] Føllesdal, D.: 1999, 'Gödel and Husserl', in: [Petitot et alii, 1999], pp. 385–400.
[Fraenkel et alii, 1973] Fraenkel, A.A. & Bar-Hillel, Y. & Levy, A.: 1973, *Foundations of Set Theory*, Second Revised Edition, North-Holland, Amsterdam, New York, Oxford.
[Frege, 1884] Frege, G.: 1884 (1989), *The Foundations of Arithmetic*, transl. by J.L. Austin, Second Revised Edition, B. Blackwell, Oxford.
[Frege, 1977] Frege, G.: 1977, *Grundgesetze der Arithmetik*, **vol. II**, §§86-137, transl. by Max Black in [Geach & Black, 1952], pp. 182-233.
[Frege, 1977a] Frege, G.: 1977, *Logical Investigations*, P.T. Geach editor, B. Blackwell, Oxford.
[Frege 1977b,] Frege, G.: 1977, 'Thoughts', in [Frege, 1977a], pp. 1–30.
[Freudenthal, 1971] Freudenthal, H.: 1971, 'Did Cauchy plagirize Bolzano?', *Archive for History of Exact Sciences*, **vol. 7**, pp. 375-392.
[Friedman, 1971] Friedman, H.: 1971, 'Higher set theory and mathematical practice', *Annals of Mathematical Logic*, **2**, pp. 325–357.

BIBLIOGRAPHY 261

[Galileo, 1638] Galilei, G.: 1638, *Discorsi e dimostrazioni matematiche intorno a due nuove scienze* in [Galileo , 844].
[Galileo , 844] Galilei, G.: 1844, *Opere Complete di Galileo Galilei*, Societá Editrice Fiorentina, Firenze.
[Gauss, 1863] Gauss, C. F.: 1863, *Werke*, **12 vols.**, Königliche Gesellschaft der Wissenschaften, Göttingen, Leipzig, and Berlin.
[Geach & Black, 1952] Geach, P.T. & Black, M. (eds): 1952, *Translations from the Philosophical Writings of Gottlob Frege*, B. Blackwell, Oxford.
[Gillies, 1982] Gillies, D.: 1982, *Frege, Dedekind, and Peano on the Foundations of Arithmetic*, Van Gorcum, Assen, The Netherlands.
[Gillies, 1995] Gillies, D. (ed.): 1995, *Revolutions in Mathematics*, Clarendon Press, Oxford.
[Gillies, 2000] Gillies, D.: 2000, 'An Empiricist Philosophy of Mathematics and its Implications for the History of Mathematics', in [Grosholz & Breger, 2000], pp. 41-57.
[Giorello, 1995] Giorello, G.: 1995, 'The "fine structure" of mathematical revolutions: metaphysics, legitimacy, and rigour. The case of the calculus from Newton to Berkeley and Maclaurin', in [Gillies, 1995], pp. 134-68.
[Glas, 1995] Glas, E.: 1995, 'Kuhn, Lakatos, and the Image of Mathematics', *Philosophia Mathematica*, (3), **vol. 3**, pp. 225–247.
[Gödel, 1933e] Gödel, K.: 1933, 'On intuitionistic arithmetic and number theory', in [Gödel, 1986-2003], **vol. I**, pp. 287–295.
[Gödel, 1951] Gödel, K.: 1951, 'Some Basic Theorems on the Foundations of Mathematics and Their Implications', Gibbs lecture, in [Gödel, 1986-2003], **vol. III**, pp. 304-323.
[Gödel, 1953] Gödel, K.: 1953, 'Is Mathematics Syntax of Language?' (*1953/9 - V), in: [Gödel, 1986-2003], **vol. III**, pp. 334-362.
[Gödel, 1961] Gödel, K.: 1961, 'The modern development of the foundations of mathematics in the light of philosophy' (*1961/?), in: [Gödel, 1986-2003], **vol. III**, pp. 375-387.
[Gödel, 1964] Gödel, K.: 1964, 'What is Cantor's continuum problem?', in [Gödel, 1986-2003], **vol. II**, pp. 254–270.
[Gödel, 1970] Gödel, K.: 1953, 'Some considerations leading to the probable conclusion that the true power of the continuum is $\aleph_2$', in: [Gödel, 1986-2003], **vol. III**, pp. 420-422.
[Gödel, 1986-2003] Gödel, K.: 1986-2003, *Collected Works*, S. Feferman *et alii* (eds.), Clarendon Press, Oxford.
[Goodman & Myhill, 1978] Goodman, N. D. & Myhill, J.: 1978, 'Choice implies excluded middle', *Zeitschrift für mathematische Logik und Grundlagen der Mathematik* 24, 461.
[Grabiner, 1981] Grabiner, J.V.: 1981, *The origins of Cauchy's rigorous calculus*, M.I.T. Press, Cambridge, Massachusetts.
[Grabiner, 1984] Grabiner, J.: 1984, 'Cauchy and Bolzano', in [Mendelsohn, 1984], pp. 105–124.
[Grattan-Guinness, 1979] Grattan-Guinness, I.: 1979, 'Bolzano, Cauchy and the New Analysis of the Early Nineteenth Century', *Archive for the History of the Exact Sciences*, **vol. 6**, pp. 372–400.
[Grattan-Guinness, 1980] Grattan-Guinness, I. (ed.): 1980, *From the Calculus to Set Theory 1630–1910*, Duckworth, London.
[Greenberg, 1980] Greenberg, M.J.: 1980, *Euclidean and Non-Euclidean Geometries*, W.H. Freeman.
[Grosholz & Breger, 2000] Grosholz E. & Breger H. (eds.): 2000, *The Growth of mathematical knowledge*, Kluwer Academic Publishers, Dordrecht.
[Hallett, 1979a] Hallett, M.: 1979, 'Towards a Theory of Mathematical Research Programmes (I)', *British Journal for the Philosophy of Science*, **30**, pp. 1–25.

[Hallett, 1979b] Hallett, M.: 1979, 'Towards a Theory of Mathematical Research Programmes (II)', *British Journal for the Philosophy of Science*, **30**, pp. 135–159.
[Hallett, 1984] Hallett, M.: 1984, *Cantorian set theory and limitation of size*, Clarendon Press, Oxford.
[Halmos, 1974] Halmos, P.R.: 1974, *Naive Set Theory*, Springer-Verlag, New York, Heidelberg, Berlin.
[Hart, 1998] Hart, W. D. (ed.): 1998, *The Philosophy of Mathematics*, Oxford University Press, Oxford.
[Helmholtz, 1878] Helmholtz, H. v.: 1878, 'The origin and meaning of geometrical axioms', [Ewald, 1996], **vol. II**, pp. 663-689.
[Hermes, 1965] Hermes, H.: 1965, *Enumerability, Decidability, Computability*, Springer-Verlag, Berlin, Heidelberg, New York.
[Heyting, 1966] Heyting, A.: 1966, *Intuitionism. An Introduction*, North-Holland, Amsterdam.
[Hilbert, 1926] Hilbert, D.: 1926 (1985), 'On the infinite', in [Benacerraf & Putnam, 1985], pp. 183–201.
[Hilbert, 1947] Hilbert, D.: 1947, *The Foundations of Geometry*, Transl. by E. J. Townsend, The Open Court Publishing Company, La Salle, Illinois.
[Hobson, 1921] Hobson, E. W.: 1921, *The Theory of Functions of a Real Variable and the Theory of Fourier's Series*, Second Edition, Cambridge at the University Press, Cambridge.
[Hoffman, 2004] Hoffman, S.: 2004, 'Kitcher, Ideal Agents, and Fictionalism', *Philosophia Mathematica*, (3), **vol. 12**, pp. 3–17.
[Hume, 1739] Hume, D.: 1739 (1983), *A Treatise of Human Nature*, Second Edition, edited by L.A. Selby-Bigge, revised by P. H. Nidditch, Clarendon Press, Oxford.
[Husserl, 1998] Husserl, E.: 1998, *Ideas pertaining to a pure phenomenology and to a phenomenological philosophy*, Transl. by F. Kersten, Kluwer Academic Publishers, Dordrecht.
[Jech, 1973] Jech, T. J.: 1973, *The Axiom of Choice*, North-Holland, Amsterdam.
[Kalmár, 1967] Kalmár, L.: 1967, 'Foundations of mathematics: Whither now?', in [Lakatos, 1967], pp. 187-194.
[Kant, 1787] Kant, I.: 1787 (1990), *Critique of Pure Reason*, transl. by Norman Kemp Smith, Macmillan, London.
[Kelly & Matthews, 1981] Kelly, P. & Matthews, G.: 1981, *The Non-Euclidean, Hyperbolic plane*, Springer-Verlag, New York, Berlin, Heidelberg, Tokyo.
[Kitcher, 1984] Kitcher, P.: 1984, *The Nature of mathematical knowledge*, Oxford University Press, New York, Oxford.
[Kleene, 1974] Kleene, S. C.: 1974, *Introduction to Metamathematics*, Wolters-Noordhoff Publishing and North-Holland Publishing Company, Groningen, Amsterdam, Oxford.
[Kline, 1972] Kline, M.: 1972, *Mathematical Thought from Ancient to Modern Times*, Oxford University Press, New York.
[Koetsier, 1991] Koetsier, T.: 1991, *Lakatos' Philosophy of Mathematics. A Historical Approach*, North-Hollland, Amsterdam.
[Kripke, 1980] Kripke, S.: 1980, *Naming and Necessity*, Blackwell.
[Kuhn, 1970] Kuhn, T. S.: 1970, *The Structure of Scientific Revolutions*, Second Edition, Enlarged, The University of Chicago Press, Chicago.
[Lakatos, 1962] Lakatos, I.: 1962, 'Infinite regress and foundations of mathematics', in [Lakatos, 1983], **vol. II**, pp. 3-23.
[Lakatos, 1963-4] Lakatos, I.: 1963-4. 'Proofs and Refutations', *British Journal for the Philosophy of Science*, **vol. 14**, pp. 1-25, 120-139, 221-243, 296, 342.
[Lakatos, 1967] Lakatos, I. (ed.): 1967, *Problems in the Philosophy of Mathematics*, North-Holland, Amsterdam.
[Lakatos, 1967a] Lakatos, I.: 'A Renaissance of empiricism in the recent philosophy of mathematics?', in [Lakatos, 1983], **vol. II**, pp. 24-42.

[Lakatos, 1971] Lakatos, I.: 1971, 'History of science and its rational reconstructions', in [Lakatos, 1983], **vol. I**, pp. 102-138.
[Lakatos, 1983] Lakatos, I.: 1983, *Philosophical Papers*, Cambridge University Press, Cambridge.
[Lakatos, 1983a] Lakatos, I.: 1983, 'Falsification and the Methodology of Research Programmes', in [Lakatos, 1983], **vol. I**, pp. 8–101.
[Lakatos & Zahar, 1983] Lakatos, I. & Zahar, E.: 1983, 'Why did Copernicus's research programme supersede Ptolemy's?', in [Lakatos, 1983], **vol. I**, pp. 168–192.
[Maddy, 1998] Maddy, P.: 1998, 'How to be naturalist about mathematics', in [Dales & Oliveri, 1998], pp. 161-180.
[Maddy, 2000] Maddy, P.: 2000, *Naturalism in Mathematics*, Clarendon Press Oxford.
[Malament, 1982] Malament, D.: 1982, 'Review of *Science without numbers: A defense of Nominalism*', *The Journal of Philosophy*, **LXXIX (9)**, pp. 523–534.
[Mancosu, 1996] Mancosu, P.: 1996, *Philosophy of Mathematics & Mathematical Practice in the Seventeenth Century*, Oxford University Press, Oxford.
[Mandelbrot, 1987] Mandelbrot, B.B.: 1987, *Gli oggetti frattali*, Giulio Einaudi Editore, Torino.
[Markov, 1962] Markov, A. A.: 1962, 'On constructive mathematics' (Russian), *Tr. Mat. Inst. Steklov.* **67**, pp. 8-14; transl. in *Amer. Math. Soc., Transl.*, II Ser. **98**, pp. 1-9.
[Martin, 1975] Martin, D. A.: 1975, 'Borel determinacy', *Annals of Mathematics*, **102**, pp. 363–371.
[Martin-Löf, 1994] Martin-Löf, P.: 1994, 'Verificationism then and now', talk given at the conference *The Foundational Debate: Complexity and Constructivity in Mathematics and Physics*, Vienna, 15-17 1994.
[Martino, 2004] Martino, E.: 2004, 'Wolves, sheep and logic', in [Assenza et alii, 2004], pp. 75–104.
[McDowell, 1994] McDowell, J.: 1994, *Mind and World*, Harvard University Press, Cambridge, Massachusetts.
[McGuinness & Oliveri, 1994] McGuinness, B.F. & Oliveri, G.: 1994, *The Philosophy of Michael Dummett*, Kluwer Academic Publishers, Dordrecht.
[Mendelsohn, 1984] Mendelsohn, E. (ed:): 1984, *Transformation and Tradition in Science: Essays in Honor of I. Bernard Cohen*, Cambridge University Press, Cambridge.
[Mendelson, 1987] Mendelson, E.: 1987, *Introduction to mathematical logic*, Third Edition, The Wadsworth & Brooks/Cole mathematics series, Belmont, California.
[Micheli, 1998] Micheli, G.: 1998, *Matematica e metafisica in Kant*, CLEUP, Padova.
[Moore, 1922] Moore, G. E.: 1922, 'External and Internal Relations', in: [Baldwin, 1993], pp. 79–105.
[Moore, 1982] Moore, G. H.: 1982, *Zermelo's Axiom of Choice*, Springer-Verlag, New York, Heidelberg, Berlin.
[Morrow, 1970] Morrow, G. R.: 1970, *Proclus A Commentary on the First Book of Euclid's Elements*, Princeton University Press, Princeton, New Jersey.
[Oliveri, 1997a] Oliveri, G.: 1997, 'Mathematics. A Science of Patterns?', *Synthese*, **vol. 112**, issue 3, pp. 379–402.
[Oliveri, 1997b] Oliveri, G.: 1997, 'Criticism and Growth of mathematical knowledge', *Philosophia Mathematica*, (3), **vol. 5**, pp. 228–249.
[Oliveri, 2005] Oliveri, G.: 2005, 'Do We Really Need Axioms in Mathematics?', in [Cellucci & Gillies, 2005], pp. 119–135.
[Palmer, 1999] Palmer, S. E., 1999, *Vision Science*, The MIT Press, Cambridge, Massachusetts.
[Parsons, 1980] Parsons, C.: 1980, 'Mathematical Intuition', in [Hart, 1998], pp. 95–113.
[Parsons, 2000] Parsons, C.: 2000, 'Reason and Intuition', *Synthese*, **vol. 125**, pp. 299–315.
[Petitot et alii, 1999] Petitot J. & Varela F. J. & Pachoud B. & Roy J-M (eds.): 1999, *Naturalizing Phenomenology*, Stanford University Press, Stanford, California.
[Popper, 1994] Popper, K. R.: 1994, *The myth of the framework*, Routledge, London.

[Prawitz, 1998] Prawitz, D.: 1998, 'Truth and objectivity from a verificationist point of view', in [Dales & Oliveri, 1998], pp. 41-51.
[Proclus, *A Commentary*] Proclus: *A Commentary on The First Book of Euclid's Elements*, in [Morrow, 1970], pp. 3–343.
[Putnam, 1967] Putnam, H.: 1967, 'Mathematics without foundations', in [Benacerraf & Putnam, 1985], pp. 295-311.
[Putnam, 1992] Putnam, H.: 1992, *Representation and Reality*, The M.I.T. Press, Cambridge, Massachusetts.
[Quine, 1963] Quine, W. V.: 1963, 'Carnap and logial truth', in [Schilpp, 1963], pp. 385-406.
[Quine, 1969] Quine, W.V.: 1969, 'Ontological Relativity', in [Quine, 1969a], pp. 26–68.
[Quine, 1969a] Quine, W.V.: 1969, *Ontological Relativity and Other Essays*, Columbia University Press, New York.
[Quine, 1981a] Quine, W. V.: 1981, 'Things and Their Place in Theories', in [Quine, 1981b], pp. 1-23.
[Quine, 1981b] Quine, W. V.: 1981, *Theories and Things*, Harvard University Press, Cambridge, Massachusetts .
[Ramsey, 1925] Ramsey, F. P.: 1925, 'The Foundations of Mathematics', in [Ramsey, 1978], pp. 152–212.
[Ramsey, 1978] Ramsey, F. P.: 1978, *Foundations*, Routledge & Kegan Paul, London.
[Rasmussen & Ravnkilde, 1982] Rasmussen, S. A. & Ravnkilde, J.: 1982, 'Realism and Logic', *Synthese*, **vol. 52**, pp. 379–437.
[Resnik, 1980] Resnik, M. D.: 1980, *Frege and the Philosophy of Mathematics*, Cornell University Press, Ithaca and London.
[Resnik, 1981] Resnik, M. D.: 1981, 'Mathematics as a Science of Patterns: Ontology and Reference', *Noūs* **XV**, pp. 529-550.
[Resnik, 2001] Resnik, M. D.: 2001, *Mathematics as a Science of Patterns*, Clarendon Press, Oxford.
[Richards, 1977] Richards, J. L.: 1977, 'The Evolution of Empiricism', *British Journal for the Philosophy of Science*, **vol. 28**, pp. 235–253.
[Richards, 1988] Richards, J. L.: 1988, *Mathematical Visions*, Academic Press Inc., San Diego.
[Rubin & Rubin, 1963] Rubin, H. & Rubin, J. E.: 1963, *Equivalents of the Axiom of Choice*, North-Holland, Amsterdam.
[Russell, 1993] Russell, B.: 1993, *Introduction to Mathematical Philosophy*, Routledge, London and New York.
[Schilpp, 1963] Schilpp, P. A. (ed.): 1963, *The Philosophy of Rudolf Carnap*, Open Court, La Salle, Illinois.
[Schirn, 1998] Schirn, M. (ed.): 1998, *The Philosophy of Mathematics Today*, Clarendon Press, Oxford.
[Schlick, 1930] Schlick, M.: 1930, 'Die Wende der Philosophie', *Erkenntnis*, **I**, pp. 4–11.
[Shapiro, 1997] Shapiro, S.: 1997, *Philosophy of Mathematics. Structure and ontology*, Oxford University Press, Oxford.
[Spivak, 1992] Spivak, M.: 1992, *Calculus*, W. A. Benjamin Inc., London, Menlo Park, California.
[Strawson, 1979] Strawson, P. F.: 1979, *Individuals*, Methuen, London.
[Szabo, 1967] Szabó, Á.: 1967, 'Greek Dialectic and Euclid's Axiomatic', in: [Lakatos, 1967], pp. 1–8.
[Taylor, 1987] Taylor, B. M. (ed.): 1987, *Michael Dummett, Contributions to Philosophy*, Nijhoff, Dordrecht.
[Tonelli,] L. Tonelli, *Lezioni di Analisi matematica*, Pisa.
[Troelstra & van Dalen, 1988] Troelstra, A.S. & van Dalen, D.: 1988, *Constructivism in Mathematics*, North-Holland, Amsterdam, New York, Oxford, Tokio.
[Tuller, 1967] Tuller, A.: 1967, *A modern introduction to Geometries*, Van Nostrand, Princeton, N.J.

[Tymoczko, 1986] Tymoczko, T. (ed.): 1986, *New Directions in the Philosophy of Mathematics*, Birkhäuser, Boston.
[Ullman, 1997] Ullman, S.: 1997, *High-level vision*, The MIT Press, Cambridge, Massachusetts.
[van Dantzig, 1956] van Dantzig, D.: 1956, 'Is $10^{10^{10}}$ a finite number?', *Dialectica* **9**, pp. 273-277.
[van Heijenoort, 1967] van Heijenoort, J.: 1967, *From Frege to Gödel*, Harvard University Press, Cambridge, Massachusetts.
[van Stigt, 1990] van Stigt, W. P.: 1990, *Brouwer's Intuitionism*, North-Holland, Amsterdam.
[Wagon, 1985] Wagon, S.: 1985, *The Banach-Tarski Paradox*, Cambridge University Press, Cambridge.
[Whitehead & Russell, 1962] Whitehead, A. N. & Russell, B.: 1962, *Principia Mathematica*, Second Edition to *56, Cambridge University Press, Cambridge.
[Wittgenstein, 1979a] Wittgenstein, L.: 1979, *Notebooks 1914-1916*, Second Edition, edited by G. H. von Wright and G. E. M. Anscombe, B. Blackwell, Oxford.
[Wittgenstein, 1979b] Wittgenstein, L.: 1979, *Notes dictated to G. E. Moore in Norway*, in [Wittgenstein, 1979a], Appendix II, pp. 108-119.
[Wittgenstein, 1981] Wittgenstein, L.: 1981, *Tractatus Logico-Philosophicus*, Routledge & Kegan Paul, London and Henley.
[Wittgenstein, 1983] Wittgenstein, L.: 1983, *Philosophical Investigations*, B. Blackwell, Oxford.
[Wright, 1983] Wright, C.: 1983, *Frege's Conception of Numbers as Objects*, Aberdeen University Press, Aberdeen.
[Wright, 1993] Wright, C.: 1993, *Realism, Meaning & Truth*, Second Edition, Blackwell Publishers, Oxford.
[Wright, 1993a] Wright, C.: 1993, 'Anti-realism and Revisionism', in [Wright, 1993], pp. 433-457.
[Wright, 1993b] Wright, C.: 1993, 'Realism, Bivalence and Classical Logic', in [Wright, 1993], pp. 458-478.
[Wright, 1998] Wright, C.: 1998, 'Response to Dummett', in [Schirn, 1998], pp. 389-405.
[Zermelo, 1904] Zermelo, E.: 1904, 'Beweis, dass jede Menge wohlgeordnet werden kann', (Aus einem an Herrn Hilbert gerichteten Briefe), *Mathematische Annalen*, **vol. 59**, pp. 514-516. Transl. in [van Heijenoort, 1967], pp. 139-141.
[Zermelo, 1908] Zermelo, E.: 1908, 'Investigations in the foundations of set theory I', in [van Heijenoort, 1967], pp. 199-215.
[Zheng, 1997] Zheng, Y.: 1997, 'The Revolution in the Philosophy of Mathematics', *Logique & Analyse*, **vol. 158**, pp. 155-173.

INDEX

abstraction, 58, 84, 102, 104, 108, 119, 123, 124, 143, 185, 211, 219
accessibility conditions, 100
activity
    creative –, 5, 29, 110, 141–144, 147
    mathematical –, 4, 15, 24, 27, 54, 74, 93, 95, 130, 147, 155, 156, 185, 186, 252
algorithm
    normal –, 153
analysis
    classical –, 145, 219
    intuitionistic –, 21, 145, 219
    Newtonian-Leibnizian –, 225
    non-standard –, 16
analytical philosophy, x, 32, 33, 102, 178
analytic/synthetic
    – distinction, 9, 10, 215
anti-objectivism, 137, 142
applicability
    – of mathematics, 61, 81, 120
a priori
    – construction
Archimedes, 74
Aristotle, 79–86, 207–209
arithmetic
    laws of –, 96, 103, 104, 121, 199
    Peano –, 94
aspect, xi, 169, 180, 184, 186, 188, 190
    – of –, xi, 163, 165

    –seeing, 176, 179, 182, 183, 186, 191, 193
    dawning of an –, xi, 166, 169, 170, 180, 185, 188, 190
assertability
    warranted –, 173, 175
attributes, 80, 82–85, 93, 101, 190, 191, 194
auxiliary hypotheses, 215, 237, 238, 240
axiom
    – elimination, 32, 43, 46
    – introduction, 32, 43, 49
    categorical – system, 122
    Euclid's –s, 55, 211
    Peano –s, 94, 103–105, 132
axiomatizable
    finitely –, 109, 114

Balaguer M., 21, 117, 118
Beltrami E., 56, 57
Benacerraf P., 98, 99
Berkeley G., 13, 14, 35, 260
Bishop E., 13, 29, 51, 139, 155–157, 251
Bivalence
    principle of –, 18, 26, 137
Bolyai J., 56
Bolzano B., 225, 228
Bombelli R., 53
Boolos G., 102
Bridges D., 20, 139, 140
Brouwer L. E. J., 19, 20, 26, 29, 92, 93, 110, 139, 142–148, 155, 157, 159, 190, 252,

257, 264
Budd M., 179, 180
Burgess J., 136

Cantor G., 27, 43, 45, 49, 51, 58, 211, 224, 225, 227–242, 244
Carnap R., 7–10, 15, 17, 29, 31, 32, 139, 263
Cauchy A. L., 205, 225, 226
cause
    efficient –, 84
Church A., 110, 112, 113, 152, 155
clarification
    conceptual –, 4
class
    absolute –es, 190
    disputed –, 18, 19, 21, 24–26, 31
    object –es, 190
    proper –es, 236
    relative –es, 190
community
    mathematical –, xi, 15, 16, 30–32, 44, 45, 52, 53, 57, 58, 75, 122, 133, 200, 201, 214, 237, 254
completeness, 4, 211
concept
    limiting –, 87
consciousness, 106–108, 112, 113, 178, 184
consistency
    absolute –, 85, 131
    – proof, 145
    relative –, 45, 56
constants
    dimensionless –, 73
    logical –, 19, 20, 140, 144, 251, 257
    physical –, 73, 160
constructibility

actual –, 160
construction
    a priori –, 90
    empirical –, 91
    pure (or schematic) –, 91
Constructivism, xi, 139
    social –, 217
content
    cognitive –, 8
    conceptual –, 116
context
    – of mathematical discovery, 110
    – principle, 101
    propositional –, 101
convention, 5, 22, 28, 51, 92, 111, 131, 148, 182, 208, 217, 237, 250
creative subject, 146, 147
criteria
    public – of correctness, 148
    rational –, xi, 15, 31, 42
Curry H., 28

decidable, 109, 114
Dedekind R., 47, 51, 103, 122, 128, 132, 219, 235, 242
definition
    explicit –, 102, 103
    implicit –, 72, 123–125, 132, 210, 211
Descartes R., 209
Dirichlet L., 226, 227
Double Negation elimination
    law of –, 19
Dummett M. A. E., x, 5, 7, 17–24, 26, 29, 31, 32, 102–104, 137, 138, 148–152, 252, 253, 259, 260, 262–264

Einstein A., 73, 74, 192
elucidation, 112, 114, 176, 220

INDEX

empiricism, 139, 197, 201
epistemology, 130, 160, 170, 184, 186
    mathematical –, ix, 33, 37
Ernest P., 217
essence
    mathematical –s, 106
Euclid, 27, 40, 51, 52, 57, 193, 194, 206–208, 211, 212, 214
Eulenberg A., 234
Excluded Middle
    law of –, 18, 19, 26, 152
existence
    ante rem –, 70
    in re –, 70
    external question of –, 8, 10, 29
    internal questions of –, 7, 9
    mathematical – assumptions, 74, 75
    physical – assumptions, 75
experience, 6, 81, 88–92, 94, 95, 107, 108, 128, 156, 164, 174, 177–179, 181, 183, 190, 191, 196–202, 209

fallibility, 2, 204, 219–221
fallible
    strongly –, 205
    weakly –, 219
falsifiability, 206, 216
    naïve –, 216
    sophisticated –, 206, 216
falsifier
    heuristic –, 205, 206
    potential –s, 204, 205
fictionalism, 135–137, 250
Field H., x, 63–73, 133, 135–138, 142, 250, 251
finitism
    strict –, 158, 160

finitists, 14, 25, 30
form
    logical –, 129, 167, 197
formal
    – conditions, 90
    – ontology, 106
    – relations, 167
    – system, 22, 23, 28, 104, 110, 121, 130, 131, 133, 145, 204, 205, 212, 224
formalism, ix
    empirical –, 28
    global –, 130, 131
foundationalism, 196, 220, 222
foundations
    – of mathematics, 4, 48, 159, 195, 196, 202, 203, 212, 220, 250, 255, 258, 261, 262
Føllesdal D., 113, 114
Fraenkel A. A., 27, 45, 49, 224
framework
    constructive –s, 155
    – for analysis, 16
    – of numbers, 7, 8
    incommensurable –s, 10, 11
    linguistic –, 7
    Newtonian-Leibnizian –, 14
    rules of the –, 9
    set-theoretical –, 41
    strict finitist –, 160
Frege G., 4, 32, 33, 40, 49, 51, 52, 61, 62, 70, 77, 81, 89, 93, 95–97, 100–106, 108, 112, 113, 120, 128, 130, 156, 165, 172, 184, 190, 196–200, 210, 211, 220, 259, 260, 263, 264

Galilei G., 74, 80, 81, 179, 193, 194
Gauss C. F., 51, 53, 54

geometry
    absolute –, 55, 56, 158
    elliptic –, 52, 56, 192
    Euclidean –, xi, 23, 40, 55, 64, 81, 95, 96, 122, 130, 187, 192, 195, 206, 208, 210–214, 223, 258
    fractal –, 82
    hyperbolic –, 23, 45, 55, 56
    non-Euclidean –, 27, 40, 52, 56
Gillies D., 85, 86, 195, 198–202
Giorello G., 13
Glas E., 217
Goodman N. D., 140
Gödel K., 22, 27, 42, 45, 85, 104, 111–114, 118, 128, 131, 145, 153, 249
grammar
    context-free –, 153
Grattan-Guinness I., 225

Hallett M. 230, 233–236
Hankel H.,226
hard core, xi, 214, 215, 224, 225, 227, 229, 230, 232, 233, 236, 237, 240–242, 247, 250–255
Hart, x
Heath Sir Thomas L., 207
Helmholtz H. v., 96, 181, 182
Hersh R., 128
heuristic, 247
    negative –, 214, 215, 241, 249
    positive –, 214, 215, 238, 240, 242, 243, 251, 253, 254
Heyting A., 19, 144, 257
Hilbert D., 29, 64, 65, 78, 84, 86, 122, 127–129, 144, 145, 157, 191, 196, 197, 202, 210–214, 220, 221, 260, 261, 264

Hilbert's Programme, 1, 2, 127–129, 144, 196, 197, 202, 220, 221
history
    external –, 35
    internal –, 34–37
Hobson E. W., 226, 227
Hoffman S., 136
holism, 8, 62
Hume D., 97, 102, 105, 196, 197
Husserl E., 105–111, 113, 114, 116, 172, 185

idealism, 114, 182
idealization, 76, 77
identity
    definition of –, 47, 48
    – conditions, 49, 50, 98, 101, 102, 105, 122, 237, 243
induction, 95, 112, 156, 198, 199
infinitesimalists, 14, 30
infinitesimals
    existence of –, 11, 12, 14, 30
    Newtonian-Leibnizian –, 218
    theory of –, 16, 33, 34
infinity
    actual –, 84–86, 93, 98, 141, 194, 226, 227, 229, 232–234, 236, 241, 242
    potential –, 141, 194, 225, 233, 242
interpretation
    intended –, 52, 100, 132
intuition
    geometrical –, 11, 51, 212
    intellectual –, 86
    – of an essence, 105, 185
    – of types, 114, 115
    mathematical –, 112–117, 185
    object of –, 100, 115, 116
    pure –, 86, 90, 91, 96, 177, 183, 185

INDEX

sensible –, 86, 91, 177
Intuitionism, ix, xi, 1, 4, 20, 26, 92, 142–149, 151–153, 197, 202, 220, 221, 251, 252, 255, 258, 260, 261, 264

justification
    coherentist –, 42

Kalmár L., 196, 197
Kant I., 24, 31, 52, 56, 79, 80, 86–96, 114, 128, 139, 142, 143, 155, 156, 171, 172, 176–178, 196, 197, 202, 209
Kitcher P., 136, 195, 198–202
Klein F., 56, 66
knowledge
    concept of –, 4
    growth of –, 213
    mathematical –, 1, 4, 5, 31, 33, 34, 36, 37, 39, 79, 81, 90, 94, 110, 111, 130, 141, 163, 191, 195, 196, 200–202, 204, 213, 217, 220, 241, 254, 255, 261–263
Koetsier T., 218, 241
Kripke S., 167
Kronecker L., 29, 139, 219
Kuhn T. S., 179, 237

Lakatos I., x, xi, 34, 36, 202–206, 214–218, 222, 237, 246, 249, 251, 253, 254
language
    first-order –, 21, 22
    logical analysis of –, 139
    mathematical –, 80–82
    natural –, 115
    private – argument, 148
    second-order –, 66

Leibniz G. W., 13, 14, 16, 35, 225
L'Hôpital G. de, 14
limitation of size
    doctrine of –, 236, 241, 243, 245, 248
Lobatchevsky N. I., 56
logic
    classical –, 18, 20–22, 130, 137, 138, 141, 149–151, 251
    first-order –, 16, 78, 144, 247
    formal –, 107
    higher-order –, 103
    – of mathematical discovery, 35, 204
    – of vagueness, 138
    predicate –, 4, 110
    intuitionistic –, 20–22, 141, 150, 151
    second-order –, 66, 67, 105
Logicism, ix, 1, 4, 103, 104, 197, 202, 220, 221

Maddy P., 21, 36, 62, 67
Malament D., 66–68
Mandelbrot B., 81, 82
Markov A. A., 25, 139, 152–154, 251, 252
Martin-Löf P., 252
mathematical objectivity, 42, 112, 137, 147, 160, 185
mathematical research programme (MRP), xi, 214, 217, 219, 221, 223–225, 227, 229–230, 233, 237, 240, 246, 250–255
    Cantor-Zermelo –, 224, 232, 233, 236–238, 246, 247, 249–251
    degenerative –, 253–255
    progressive –, xi, 217, 224, 246, 250, 251, 253, 254
    set theory —, 225

mathematics
    finitary –, 145
    history of –, ix, 1, 3, 4, 30,
        32–34, 40, 51–53, 60, 120,
        131–133, 204, 224, 227,
        250, 258
    informal –, 202, 205
    intuitionistic –, 21, 25, 146,
        148
    philosophy of –, ix–xi, 1, 5,
        17, 26, 31, 33, 34, 77,
        137, 146, 154, 160, 195,
        200, 206, 214
matter
    intelligible –, 84
    sensible –, 84
McDowell J., 182
meaning
    – is use, 23
    numerical –, 157
    theory of –, 19, 20, 31, 150
measurement
mental states, 97
metaphysics
    from logic to –, 22, 23, 26, 31
method
    axiomatic –, 110, 123, 130,
        131, 250
    – of verification, 139
methodology
    – of scientific research progr-
        ammes (MSRP), 36, 37,
        218
    normative –, 35–37
Micheli G., 91
Mill J. S., 33, 165, 195, 197–200,
    202
model
    criterion of the –, 52, 53
    isomorphic –s, 99, 100, 117–
        121

non-standard –, 16
Moore G. H., 43, 44, 232
Myhill J., 140

naturalism
    – about mathematics, 36
nature
    laws of –, 62, 75, 77, 192
Newton I., 12, 13, 16, 35, 65, 74,
    133, 134, 218, 225, 237,
    260
nominalism, 66, 67, 69, 134, 136,
    137
number
    cardinal –, 43, 82, 85, 98, 104,
        108, 141, 229, 231, 232,
        234, 235, 238, 240, 248
    laws of –, 62
    natural –s, 8, 28, 29, 60, 85,
        98, 100, 101, 105, 115,
        117, 119, 120, 140, 143,
        146, 152, 153, 191, 193,
        194, 198, 219

object
    abstract –, 27, 66, 70, 97, 98,
        100, 105, 107, 116, 117,
        134, 148, 152, 163
    fictional –s, 36
    macroscopic –s, 72, 192
    mathematical –s, 17, 18, 21,
        22, 36, 68, 82, 84, 91–93,
        98, 100, 113–120, 134, 135,
        137, 198, 218
    physical –s, 36, 66, 67, 76,
        191, 198
    private –s, 93
objectivity
    mathematical –, 42, 112, 137,
        147, 160, 185
observation sentence, 178
Ockham's razor, 15, 74

INDEX

Oliveri G., x, xi
ontological argument, 24
ontological commitment, 9, 10, 62, 63, 67, 74, 75, 77, 78
ontology
   formal –
   mathematical –, ix, 39, 79, 81, 84, 160
   nominalistic –, 65
ordinals
   von Neumann –, 41, 50
overview, 3–5

paradigm
   metaphysical –s, 237
paradox
   Berry's –, 46
   Burali-Forti's –, 46, 48
   Cantor's –, 46
   Grelling's –, 46
   Richard's –, 46, 48, 244
   Russell's –, 46, 48
   Tarski-Banach –, 44
Parsons C., 105, 114, 115
pattern
   mathematical –, 169, 171, 183
      cumulative hierarchy of –s, 172
   – of patterns, 123, 124, 193
   – recognition, 123, 124
   –s of points, 81
   science of –s, xi, 122, 163, 170, 184, 185, 195
   set-theoretical –, 164
Peano G., 48, 51, 122, 131
perception
   – of tokens, 114–115
phenomena, 183
   natural –, 3, 81, 179
phenomenological reduction, 108, 116, 185

phenomenology, 109, 113, 114
physics
   nominalization of –, 67
Plato, 39
Platonism, 27, 33, 69, 83, 97–100, 103, 117–121, 123, 134, 149, 150, 163, 170, 176
   full-blooded –, 21, 117
   plentiful –, 21
Poincaré H., 56
Popper K. R., 15, 204, 216
posits, 8, 63, 74, 75, 77, 78, 206, 209
practice
   mathematical –, 1, 4, 31, 33, 122, 130, 141, 205, 260
Prawitz D., 25, 26, 151, 152, 252, 253
primitive notion, 72, 98, 132, 210
probabilistic inference, 68
probability, 68, 69, 71, 72, 196, 199
procedure
   constructive –, 4, 220
   decision –, 146
   – of verification, 5, 42, 139
Proclus, 209
progression
   abstract –, 119
   ordinary –, 119
proof
   canonical –, 25
   constructive –, 18, 25, 42, 92, 93, 139, 142, 144, 147, 154, 157, 160, 251
   –theory, 129
   realm of –s, 25, 26, 152
propositional function, 47
   definit –, 246, 247
protective belt, 215, 224, 225, 237, 242, 247, 250

provability, 25, 29, 147, 150, 252, 253
pseudo-question, 8
psychologism, 93, 147, 148, 184, 185
Putnam H., 8, 9, 173–176
Pythagoras, 39

quantification
    plural –, 67
quantum
    – measurement, 6
    – mechanics, 6
Quine W. V., 9, 10, 50, 57, 58, 62, 64, 67, 75, 76, 105, 116, 118–121, 176, 259

Ramsey F. P., 48, 49, 76
realism
    anti– about truth, 28
    Aristotelian –, 79, 82, 94
    global –, 78
    internal –, 8, 9, 173
    Kantian –, 79, 86
    metaphysical –, 18, 28, 29, 31, 32, 39, 40, 138, 151
    Pythagorean –, 79, 80
    – about truth, 28, 29, 31, 138, 151
    –/anti-realism debate, 2, 3, 7, 9, 17, 20, 21, 31–33, 36, 52, 138
    set-theoretical –, 39, 41, 49, 79, 190, 224
    structural –, 1, 170, 176
reality
    ideal –, 8
    material –, 8
    mathematical –, 5, 6, 39, 40, 43, 46, 49, 51, 79, 81, 82, 84, 93, 97, 111, 112, 152, 163, 195

    set-theoretical –, 42
reconstruction
    nominalist – of physics, 133
    rational –, 34–37, 42, 215, 218, 223
reducibility, 6, 72
reference
    inscrutability of –, 76, 105, 116, 120, 176
relation
    external –s, 167
    internal –s, 166, 167
    structural –s, 167
representation
    faithful –s, 59
    mental –, 52, 56
    non-perceptual –, 171, 188
    pre-reflective –s, 112, 190
    system of –, 89, 90, 163–165, 167, 170, 171, 174, 181, 185–187, 190, 191, 193
Resnik M., 68, 69, 118–122, 124, 170
Riemann B., 18, 149, 226
rigour, 4, 14, 15, 111, 205, 260
Robinson A., 16
Russell B., 1, 40, 46–49, 76, 98, 104, 121, 134, 249

Saccheri G., 45, 55
scepticism, 1, 44, 218, 221
schema, 37, 91–93, 140, 143, 173, 175, 176, 243, 248
Schlick M., 139
science
    quasi-empirical –, 2, 195, 223
scientific research programme (SRP)
    degenerating –, 216
    progressive –, 216
seeing
    – something as, 164, 176, 180, 181, 184

– that, 124
semantics
  truth-conditional –, 149, 150
sense
  – of propositions, 101
sensibles, 83
sets
  Borel –, 248
  cumulative hierarchy of –, 245, 249
  definable –, 248, 249
  well-founded –, 249
Shapiro S., 119, 122, 124, 170
solipsism, 148
space
  perceptual –, 171, 188, 189
space-time
  continuity of –, 67
  – points, 66–68
  – regions, 67, 68
  substantival view of –, 67
statement
  basic –, 203, 204
  finitary –, 84
  ideal –s, 29, 78, 84, 85
  mathematical –, 5, 9, 21, 27, 29, 96, 137, 139, 148–150, 156, 158, 197, 200–203, 212, 222
    absolutely undecidable –s, 24, 151
  number-theoretical –s, 17, 18
  pseudo–, 139, 157
structure
  ante rem –s, 121, 171
  mathematical –, 119, 122, 124, 132, 186
  number-theoretical –, 17, 18
  science of –s, 24, 117–119, 210
substance, 83, 175
surveyability, 41, 97

syntax
  logical –, 133
Szabó Á., 207, 208

Tartaglia, 53
theory
  Cantor-Zermelo set –, 36, 37, 223, 224, 240, 250–253, 255
  computability –, 4
  Euclidean –, 65, 203
  general relativity –, 68
  informal –, 132, 205
  naïve set – (**NST**), 224, 233, 234, 236–238, 240–242
  Ramified – of Types (RTT), 46–49
  scientific –, 4, 7, 62, 63, 73–76, 135, 179, 216
  – of cumulative types
  – of types, 48
third realm, 106, 112
thought
  law of –, 102
  realm of –, 102
tolerant pluralism, 21–23
Troelstra A. S., 153, 154
truth
  determinate –value, 137, 138
  mathematical –, 4, 6, 17, 18, 21, 22, 26–29, 31, 147, 199, 212
  self-evident –, 43
Tymoczko T., x

Ullman S., 190, 191
understanding, 4, 15, 19, 20, 22, 29, 31, 35, 87, 89, 92, 97, 102, 112, 113, 115, 121, 128, 129, 140, 143, 154, 173–177, 196, 208, 212, 220, 231

universe
    'blanket' –, 234
    iterative –, 234
    – of sets, 22, 74, 241, 242
    von Neumann –, 245
use
    conditions of correct –, 85

vagueness, 138
    logic of –, 138
validity
    empirical –, 28, 156
van Dalen D., 153, 154
van Stigt W. P., 144
verifiability, 28, 84
verification independent, 152
verificationism, 6, 139, 140, 143, 146, 159, 160, 197

Weierstrass K., 16, 189, 205, 225, 226, 228
Wittgenstein L., xi, 23, 86, 148, 164, 166, 167, 169, 170, 179–183
world
    external –, 4, 62, 80, 81, 93, 97, 106, 177
    material –, 85
    physical –, 63, 64, 67, 68, 85
Wright C., 102–104

Zahar E., 216, 246, 249
Zermelo E., 44, 48, 58, 99, 224, 232, 240–243, 246–248
Zheng Y., ix

www.ingramcontent.com/pod-product-compliance
Ingram Content Group UK Ltd.
Pitfield, Milton Keynes, MK11 3LW, UK
UKHW021317180426
11947UKWH00015B/1287